UNCOVERING

THE SECRETS OF

THE RED PLANET

ARS

NATIONAL GEOGRAPHIC SOCIETY

UNCOVERING
THE SECRETS OF
THE RED PLANET
MARS

PAUL RAEBURN

FOREWORD AND COMMENTARY BY

MATT GOLOMBEK

Cutaway illustration of Mars reveals the central metallic core of the only other rapidly spinning terrestrial planet besides Earth. A 360° panorama (continuing over the following six pages) shows the Pathfinder lander, a rocky plain, streamlined hills, and the rover next to the rock Yogi.

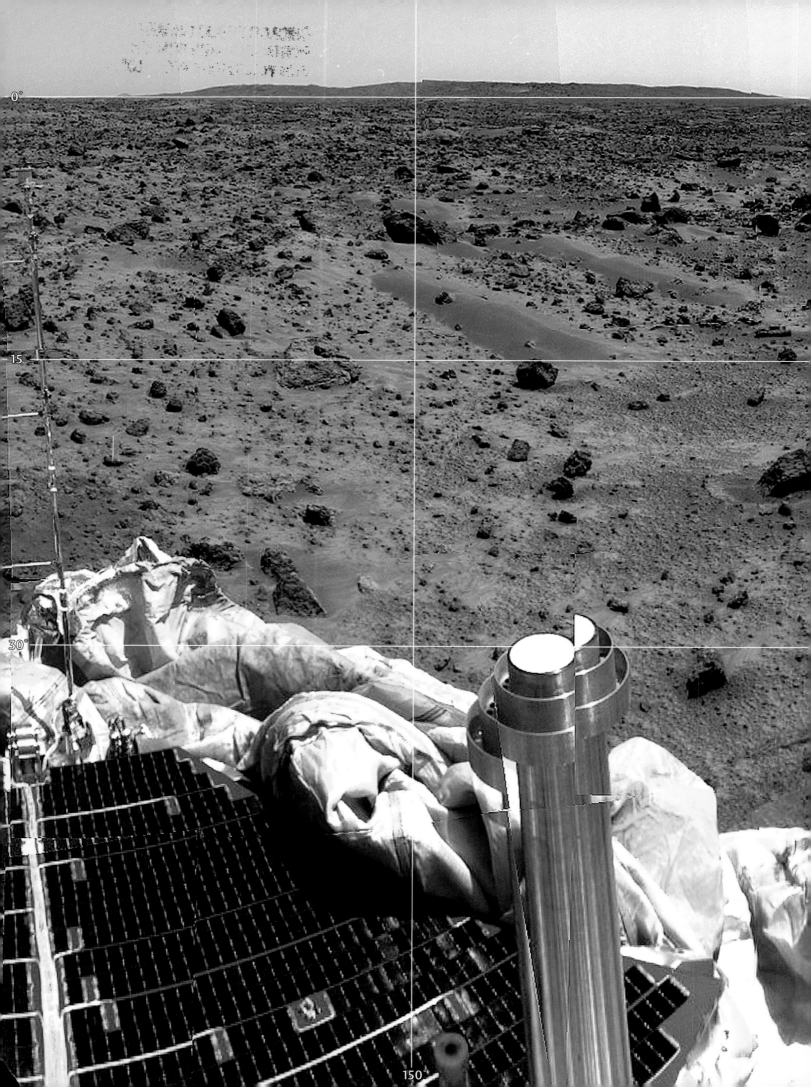

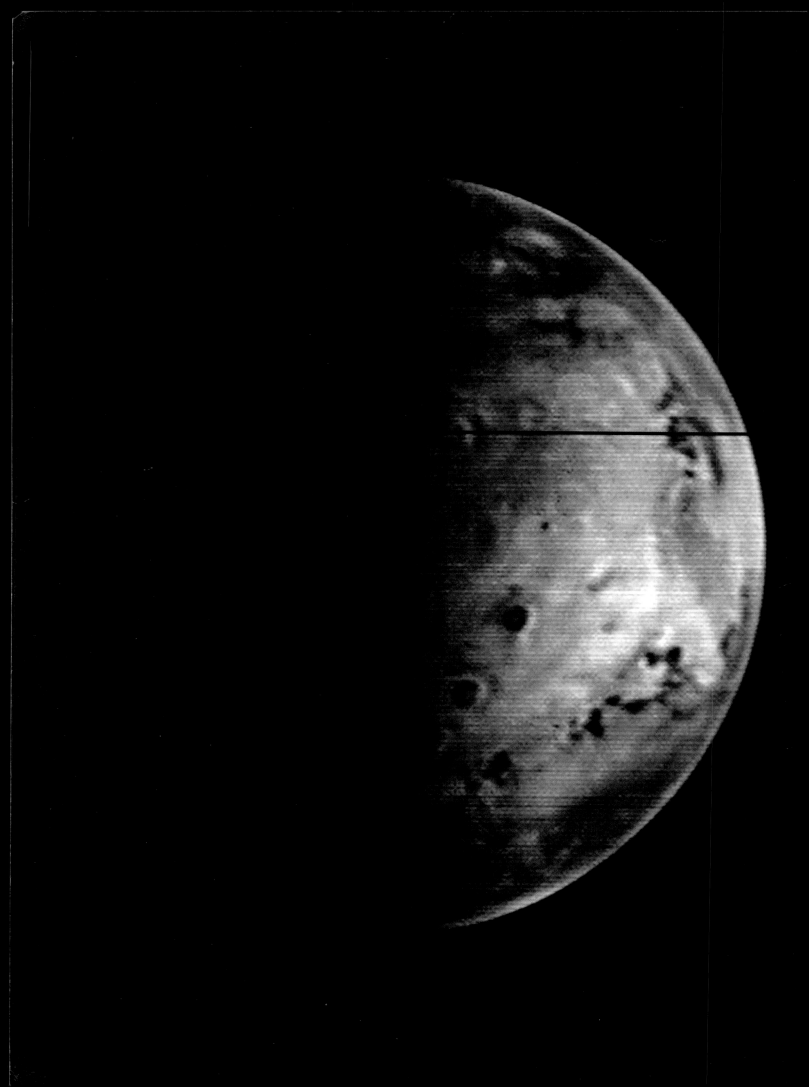

As it approached Mars on the morning of August 21, 1997, the Mars Global Surveyor shot this image, revealing Olympus Mons and three other giant volcanoes.

EXPLORING MARS

I am a research scientist who has been working on Mars for most of my professional career. Prior to July 1997, that has meant working almost exclusively with one set of 20-year-old data. In 1977, as a graduate student studying geology, I worked on the mission that returned that data set as an associate of a guest investigator on the extended Viking Orbiter mission. At that time, the Viking project office at the Jet Propulsion Laboratory in Pasadena, California, was a pretty sleepy place. The Mars landings had occurred the year before and the orbiters were in their extended missions. Most of the scientists were back at their home institutions, and the orbiters appeared to be run by a small group of engineers. I still remember watching with amazement as images of Mars, which was hundreds of millions of miles

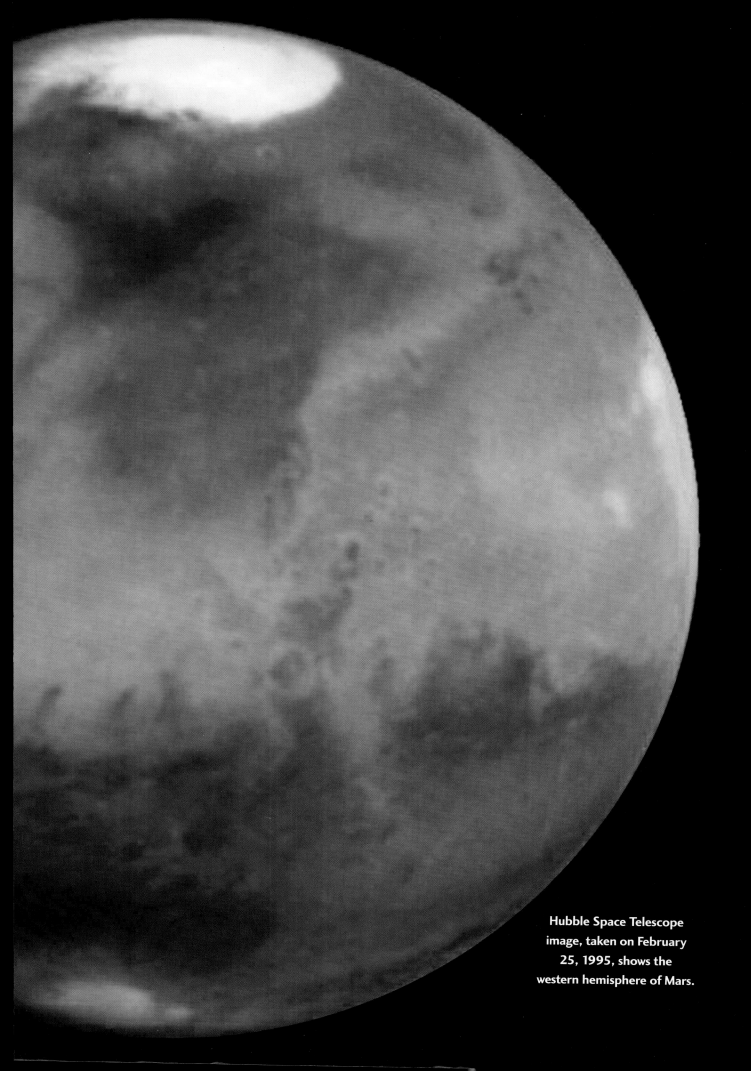

Hubble Space Telescope image, taken on February 25, 1995, shows the western hemisphere of Mars.

away, came down to Earth for the first time—each new sequence of images uncovering areas of Mars never seen before in detail.

Why would someone spend his entire career looking at data that was 20 years old? Well in 1977, the Viking images were not old and there were thousands of them—about 25,000 images of the surface from orbit. It also took a while to sort through all the data and to design and carry out research projects on particular aspects of them. By the time the data became old, it didn't matter—Mars is such an interesting place.

The fourth planet from the sun and the last terrestrial (or rocky) planet (the outer planets are gaseous), Mars is the most Earthlike planet in the solar system—it has seasons, days similar in length to the Earth's, and an atmosphere and polar caps that change with the seasons (no other planet has all these characteristics). It is the only body in the solar system, besides the Earth, where liquid water is known to have flowed across the surface, and evidence hints at an early environment in which liquid water may have been in equilibrium with the environment and stable on the surface. Liquid water is an absolute requirement for life as we know it, and the tantalizing data returned by the Viking orbiters suggests a planet in which the conditions early on in its history may have been conducive to the formation of life. We know, for example, that life started on the Earth soon after its formation. The oldest rocks on the Earth (3.9 and 3.6 billion years old) have strong chemical and clear fossil evidence of life, indicating life must have started before then. Terrains on Mars with evidence for liquid water on their surfaces (such as dry lake beds) likely date back to a similar period. By studying our neighboring planet, we can address one of the most fundamental scientific questions—are we alone in the universe? Will life form anywhere that liquid water is stable, or does it require something else as well? If life did start on Mars, what happened to it? Trying to answer questions such as these, we continue the profound drama of attempting to understand the solar system and our place in it.

Scientists generally think that the prospect for life existing on Mars today is remote. The Viking landers, which touched down on the surface of Mars in two different places, carried sophisticated life detection packages that were designed to look for metabolic activity in the soil. The instruments did not find evidence for life. Instead they found an amazingly harsh environment in which liquid water is not stable anywhere. The temperature and the atmospheric pressure on Mars are so low that water could exist only as water vapor or as ice. In addition, the surface is bathed in large doses of ultraviolet radiation, which is generally lethal to life. One Viking instrument could find no evidence of any organic (complex, carbon-based) molecules at all. This was a real surprise, given that organic molecules are common constituents of meteorites that are expected to impact the terrestrial planets regularly. It actually looked like oxidants in the surface soil were destroying all the organic molecules. Life on Mars today is unlikely. It is possible that life could exist in subsurface thermal springs or other niches, but finding such life would be difficult. We are, therefore, forced to look to the geological record to find out what the environment was like early in the planet's history and whether life could have started on Mars.

Finding out what happened on Mars early in its history was the scientific goal of the Pathfinder mission. Pathfinder landed at the mouth of a giant channel created by a catastrophic outflow of water to look for rocks carried down from the ancient highlands to the south. Given our rudimentary knowledge of what rock types make up Mars, landing at a location where a variety of rock types might be present offered the prospect of understanding what kinds of materials make up the crust of Mars. Each rock carries a story of how it formed. By examining its minerals, color, fabric, and texture, a geologist can identify the type of rock, how it formed, and the nature of the environment in which it formed. In Pathfinder's case, the rover, Sojourner, can be viewed as a diminutive one-foot geologist. By imaging the rocks up close and analyzing their chemical components, using an instrument known as an alpha-proton x-ray spectrometer, it was hoped that rock types would be identified. By landing near the mouth of an outflow channel that drains the ancient, heavily cratered terrain on Mars, it was hoped that scientists would be able to identify the processes that formed the planet's ancient crust and the nature of its early environment. Was early Mars warmer and wetter? Did Mars undergo a major change in climate (it is clearly not warm and wet at present)?

We wanted answers to these questions, and to get them we needed to locate a suitable landing site. Deciding where to land Mars Pathfinder was an arduous two-and-a-half-year effort that concentrated far more on the safety of the site than the potential science. If the team did not land Pathfinder safely, there would be no science. We did not have detailed information about what we would find at the site, and we could not predict what we would find both scientifically and in terms of potentially dangerous terrain. In order to select a site, we first had to understand the constraints that the spacecraft design and landing system and sequence placed on the landing site. These engineering constraints required that we land at a location low enough to allow the parachute ample time to bring the lander to the correct terminal velocity. The lander and rover are solar powered so that landing near the subsolar latitude was important. But most important was determining the characteristics of the terrain—characteristics that would be considered "safe" or hazardous for landing. That forced us to determine the presence of steep slopes and high scarps as well as the distribution of rocks and the amount of fine-grained dust that could be hazardous to the lander, or that might reduce the trafficability of the rover, or limit the lifetime of the solar powered lander and rover. Unfortunately we could not see any of these hazards. The highest resolution images of the landing area allowed the resolution of features that are no smaller than large sports stadiums on Earth; features potentially hazardous to the lander, on the other hand could be as small as large television sets. We had to use a wide variety of remote sensing data at much larger scale to infer the properties of the surface at a smaller scale. After two and

a half years of studying all possible locations on Mars that met the basic engineering constraints, we decided to land in Ares Vallis in the low area of Chryse Planitia. This area appeared to be a flat, relatively smooth area with little dust that appeared safe for the Pathfinder lander and offered the prospect of carrying out important scientific investigations.

Although searching for life on Mars forms the centerpiece of NASA's long-term Mars exploration program, we have learned a tremendous amount about the red planet from the Viking missions. Mars is a unique planet. In addition to being the most Earth-like terrestrial planet, Mars is intermediate in size between the large terrestrial planets (Earth and Venus) and the small terrestrial planets (Mercury and the moon). The longevity and vigor of internal geologic activity among the terrestrial planets correlates roughly with size, with the Earth and Venus being the most active and Mercury and the moon being the least active. This relation between size and geologic activity is understandable, considering that most geologic and tectonic processes are driven by the heat engine of a planet, and large objects (in this case, planets) take more time to cool and probably have had greater contributions of heating from accretion, differentiation, and radioactivity than small planets. The large terrestrial planets are active today and have continually renewed their surfaces, so most old rocks and terrains are destroyed. In contrast, the small planets have not been active for billions of years, so their surfaces are dominated by heavily cratered terrain formed by impact during the first billion years or so of solar system history. Mars is just large enough to have remained active for most of solar system history (4.6 billion years), but not so active that all record of its early history has been destroyed. This intermediate level of activity has produced rocks on the surface of Mars that uniquely record the entire history of the solar system.

A global look at Mars reveals that the surface can be divided into two or three major terrains. The southern two-thirds of the planet is composed of heavily cratered terrain, which likely dates back to the first quarter of solar system history. Basically the more craters a surface has, the older it is, and surfaces that are very old are very heavily cratered. The northern one-third of Mars is lightly cratered, indicating it is relatively young. On average the heavily cratered terrain stands about two kilometers higher than the northern lowlands. Geological relations along the boundary indicate it is old, so that some process lowered the northern one-third of the planet very early in the planet's history. One clue is that the boundary between the terrains is roughly circular. One possible explanation is that a giant impact hit the north early in the history of the planet, thereby lowering it and making a site for continued infilling by volcanics and sediments.

Standing astride the highland-lowland boundary in the western hemisphere of Mars is the Tharsis province, which is unique in the solar system. It is a bulge that rises to ten kilometers above the datum (the equivalent of sea level on Mars), with four enormous volcanoes (the largest in the solar system) and associated volcanics sitting atop. The "glue" that holds the entire area together is a system of tectonic features that cover the entire western hemisphere of Mars (one half of the planet). These features include grabens and rifts, which are long linear valleys that formed when the crust was pulled apart in extension, radiating out from the center of Tharsis. Most are relatively narrow, but the largest is enormous—Valles Marineris. Valles Marineris is a canyon system more than 2,500 miles (4,000 km) long and up to 6 miles (10 km) deep. What an unprecedented view of the geologic history of Mars would await an observer who gazed upon those cliffs, much like a geologist looking in the Grand Canyon. But Valles Marineris is six times deeper! At the western end of Valles Marineris near the top of the bulge is Noctis Labyrinthus, the largest labyrinth of intersecting canyons in the solar system. Other common tectonic features around Tharsis are the wrinkle ridges, which are predominantly concentric, and by analogy with similar ridges on Earth, are thrust faults and folds that formed when the crust was compressed or pushed together. Both systems of tectonic features appear to have resulted from adjustments in the outer layers of the planet due to the formation of the Tharsis bulge.

Valles Marineris ends on its eastern side in a series of chaotic depressions that form the upper reaches of a system of catastrophic outflow channels that drain into the northern lowlands, which in this area is named Chryse Planitia. These outflow channels were truly catastrophic events, forming when huge volumes of water (hundreds of cubic kilometers) flooded across the surface, carving each one in a couple of weeks. Both Viking 1 and Pathfinder landed in this region of Mars, because it is a very low area (two kilometers below the datum) near the equator. At Ares Vallis, in which Pathfinder landed, scientists estimate that it would have been the equivalent of taking all the water in the Great Lakes and draining it to the Gulf of Mexico in a two-week period. The canyon that would be left behind would be up to a mile deep and a couple of thousand kilometers long.

Where did all the water come from to produce the catastrophic outflow channels on Mars? At the uppermost reaches of the channels there is no evidence for lakes that were dammed. Instead it looks as if the water originated in zones of chaos, or chaotic terrain. This terrain has a disrupted uneven hummocky appearance, and it appears that the water actually came up out of the ground and flooded downhill. How did this happen? We don't know for sure, but it was probably through some combination of processes, in which the crust is heated and ice is melted (perhaps by an intrusion of molten rock), with a subsequent impact or "marsquake" shaking the area and mobilizing the water to flood downhill. Most of the outflow channels form in this region of steep topographic gradient, with a difference of four kilometers between the highlands and the lowlands.

We learned a lot about catastrophic outflow channels by studying analogous features on the Earth. One that was studied to better understand the Ares Valles region and the Pathfinder landing site was the Channeled Scabland of Washington State. At the end of the last ice age, about 13,000 years ago, a lobe of the Cordilleran ice sheet dammed a glacial lake called

Lake Missoula in the region of present-day Lake Pend Oreille in Idaho. Glacial Lake Missoula, which covered western Montana, contained roughly the water now held in Lakes Erie and Ontario combined. The ice dam ruptured, and the water in the lake rushed to the Pacific, carving a series of channels and stream-lined islands or hills now known as the Channeled Scabland. Channelized water flowing down the Grand Coulee debouched into the Quincy Basin at Soap Lake and filled the depression. When the water reached the basin, its velocity slowed and the sediment it was carrying was deposited into a 40-kilometer-long fan, known as the Ephrata Fan. The largest boulders were deposited near Soap Lake, with smaller particles falling out of suspension farther down the fan where finer sand and gravel were deposited near the end of the fan near Moses Lake, Washington. We investigated the upper reaches of the Ephrata Fan and found some very rocky surfaces where late-stage drainage of the basin (after the sediment had been deposited) carried away the finer sediment in small channels. Other areas nearby showed moderately rocky surfaces (20 percent of the surface covered by rocks) that would be safe for landing and roving, with many rocks to investigate with our one-foot geologist rover. We suggested that because the Pathfinder landing site was located a hundred kilometers downstream from the mouth of Ares Vallis, where it debouched into Chryse Planitia, most of the largest boulders (esti-mated to be up to ten meters high) would be deposited upstream and that predominantly smaller clasts would be present at the Pathfinder site (a prediction that appears to be correct).

What happened to all the water? There are certainly no oceans on Mars now, although some investigators suggest that these outflow channels did fill the north-ern lowlands with an ocean after they flooded. Oth-ers suggest that the channels drained into the northern plains and simply froze and became buried by sub-sequent floods and/or sediments and dust. But if the early environment on Mars was warmer and wetter, what happened to all the water in a general sense—where did it go? Many scientists suggest that Mars is

a water-rich planet, but that the water is not now liquid so that it is difficult to "see." A number of reservoirs for water on Mars have been proposed. A variety of landforms in the higher latitudes of Mars suggest that there could be quite a bit of water locked up as ground ice or permafrost (up to a few kilome-ters thick), which is ice frozen beneath the ground. We know from spectroscopic measurements that the northern polar cap is composed at least partially of water ice. Very small quantities of water vapor have also been detected in the atmosphere. The ubiquitous bright red dust on Mars, which gives the planet its red color, is likely a rusty weathering product that prob-ably involved water to form and still has some locked up in it. Some water in the atmosphere could have been disassociated and/or lost to space, perhaps stripped by the solar wind. Finally, if Mars has a lot of water, there could even be a groundwater table beneath the ground ice, with liquid water filling the pore space in the rocks, similar to that here on Earth. Estimates of the total volume of water locked up in these reservoirs on Mars would be make the reser-voirs hundreds of meters to kilometers thick if made into a hypothetical global ocean on top of the surface.

If Mars was warm and wet in the past, why did it change? What happened to produce the cold and dry environment of today? Scientists do not know the answer to this question, but it is obviously impor-tant for understanding long-term climate change in general. Here on Earth, for example, there is evidence that the burning of fossil fuels may have elevated the carbon dioxide level of the atmosphere, which has contributed to global warming. How big an effect this might be having is greatly debated. Is this change small and insignificant in comparison to global changes, such as those oscillations between ice ages, for example, or is this a large effect with serious long-term repercussions? Scientists do not now know the answer to this important question. One way to help investigate it is to look into the geologic record at previous climatic changes to better understand them. Another way would be to look at another planet where a major change in climate has occurred to better understand it and why it happened. By trying to better understand the planets in our solar system, we become more informed about what processes are important on Earth and how they might change under different conditions.

Scientists also learned a lot about Mars by studying twelve extraordinary rocks on Earth that most scientists are confident have come from the red planet. These rocks are often referred to as the SNC meteorites, short for "Shergotty-Nakhla-Chassigny." Of these twelve, four were observed to fall from the sky, so we know they are not from the Earth: one in Chassigny, France; one in Nakhla, Egypt (which is reported to have killed a dog); one in Zagami, Nigeria; and one in Shergotty, India. Six were found on the Antarctica ice cap, one was found in Indiana, and another was found in Brazil. By carefully studying these rocks using very sophisticated instruments, scientists have uncovered an extraordinary story of how, when, and where these rocks formed and how and when they got to the Earth. All the rocks are similar to the most common rock type on Earth, called basalt. Basalts are fine-grained dark igneous rocks relatively rich in iron and magnesium (and consequently relatively low in silicon), which form when the top of the mantle (the dense region beneath the lighter crust) melts and that melt rises directly up through the crust and freezes on the surface. Basalts underlie most of the oceans and form at mid-oceanic ridges, where the plates on the Earth are being pulled apart and mantle melts and freezes at the surface. Oceanic crust made of basalt is relatively dense, low-standing and mostly underwater, compared to less dense, high-standing continental crust on the Earth. In addition, all of the rocks are much younger than all other meteorites, which all date back to the origin of the solar system, about 4.6 billion years ago, and are thought to be from the asteroid belt, a zone of rocky objects between Mars and Jupiter. Most of the meteorites show some evidence of being shocked in what scientists believe marks the impact event or events that blasted these rocks off of Mars and into orbit.

Radioisotope-dating techniques indicate that the shergottites (named for Shergotty) crystallized (or froze) about 180 million years ago. Cosmic-ray exposure ages show they spent up to a few million years in space after being strongly shocked, and those that were found on Antarctica have been on the Earth for a few tens of thousands to a couple of hundred million years (from surface weathering). The nakhlites and chassignites (named for Nakhla and Chassigny) crystallized about 1.3 billion years ago, have been in space for over 10 million years, and have been on the Earth for less than 200 years. They all formed in oxidizing conditions, and some formed in hydrous conditions (water present). The minerals in the rocks and the overall chemistry is not very different from basalts on the Earth, except for the ratios of certain elements. In addition, all the rocks have similar oxygen isotopes, distinct from all rocks on Earth and thus requiring that they formed on a different planet with a different evolutionary history. A couple of the meteorites have gas trapped in glassy nodules, whose composition matches that of the atmosphere of Mars as measured by the Viking landers, which has unique signatures of nitrogen, argon, zenon and oxygen. Finally, where else could these meteorites have come from? Mars is the only viable option. The planet where these rocks formed must be large enough to have volcanism throughout geologic time (which rules out Mercury and the moon) and has to have had water (which rules out Venus). No planet except Mars has all these attributes.

In 1997, scientists studying a newly identified Martian meteorite found near Allan Hills, Antarctica, called ALH 84001, announced that they had found evidence for possible life in this rock. This meteorite is older (4.5 billion years old) than the other meteorites from Mars, but shares many of the chemical characteristics discussed above. Within fractures inside the rock, researchers found very small (up to a few hundred microns in size) carbonate globules. They also found 10 to 100 micron-size crystals of magnetite and iron sulfides and organic molecules called polycyclic aromatic hydrocarbons. Finally they described very small (about a tenth of a micron in size) elongate features that bear some morphologic resemblance to bacteria microfossils on Earth. They suggested that having all four features together indicates that the small bacteria produced the carbonate, magnetite crystals, and polycyclic aromatic hydrocarbons (as occurs on Earth), suggesting for the first time that life may have begun somewhere besides the Earth. Other scientists argued that there was no direct evidence of biologic activity, that inorganic precipitation of the carbonates was a possibility, that polycyclic aromatic hydrocarbons are common in meteorites, and that the small elongate features observed are far smaller than any known lifeform and too small, in fact, to actually accommodate the molecules necessary for life. Although the possibility that ALH 84001 actually contains evidence of life will likely be argued for some time, there has also been a change in scientists' perception of acceptable niches for life on Earth. Specifically, in the 20 years since Viking, scientists have discovered that among the most primitive life on the Earth are hydrothermal organisms that live off steam and hot springs at mid-oceanic ridges (where no light penetrates). Life has also been found buried kilometers below the surface and in extreme environments that previously were thought to be inhospitable. These discoveries seem to make the idea of life starting in similar environments on Mars and perhaps hanging on in specialized environments less extreme.

So, you ask—was it worth waiting 20 years to return new data from Mars via Pathfinder to Earth? With some bias, after five years of hard work on the project, the answer is an unqualified yes. Data returned by Pathfinder have already significantly changed our understanding of Mars and will continue to be studied for years and years to come. Using the assortment of science instruments and the rover, Pathfinder was able to address a wide variety of scientific topics that include the geology and geomorphology of the surface, the mineralogy and geochemistry of rocks and soils, the physical properties of surface materials, the magnetic properties of airborne dust, the structure and meteorology of the

atmosphere (including the characteristics of dust, clouds and fogs), and the rotational and orbital dynamics of Mars.

Overall, Pathfinder found evidence for an even more Earthlike planet than scientists appreciated. The rover discovered rocks that are completely different from the Martian meteorites and generally similar in composition to the continental crust on Earth. If these rocks are representative of the ancient cratered terrain, they suggest more Earthlike processes of crustal formation on Mars than previously believed. The rover also found rocks that appear sedimentary in origin and some that may be conglomerates, which suggests that liquid water was stable for a long period of time (required for their formation), and that the climate was warmer and wetter than at present. If this hypothesis is correct, then Mars was climatically much more like the Earth in the past and therefore much more conducive to the formation of life. Pathfinder also found strong evidence for sand-size particles on Mars that typically form on the Earth by fluvial processes, in which liquid water breaks rocks into smaller and smaller pieces. If sand on Mars formed in a similar manner, then these observations could further argue for the role of water in the weathering of materials on its surface. Magnets on Pathfinder found airborne dust that is very magnetic, which can best be explained by an active hydrologic cycle on Mars in which liquid water dissolved iron out of crustal materials, which was freeze-dried onto dust grains in a highly magnetic mineral known as maghemite.

Will our ongoing exploration of Mars indicate a more Earthlike planet with a warmer and wetter past? Will the current Mars exploration program return samples to Earth that indicate life actually got started on the red planet? Could life have originated on Mars and migrated to the Earth in one of many impact exchanges between the terrestrial planets? I don't know the answer to these questions, but I do feel privileged to be able to take part in the continuing drama of exploration and discovery of Mars that so transfixed and amazed the world in July 1997.

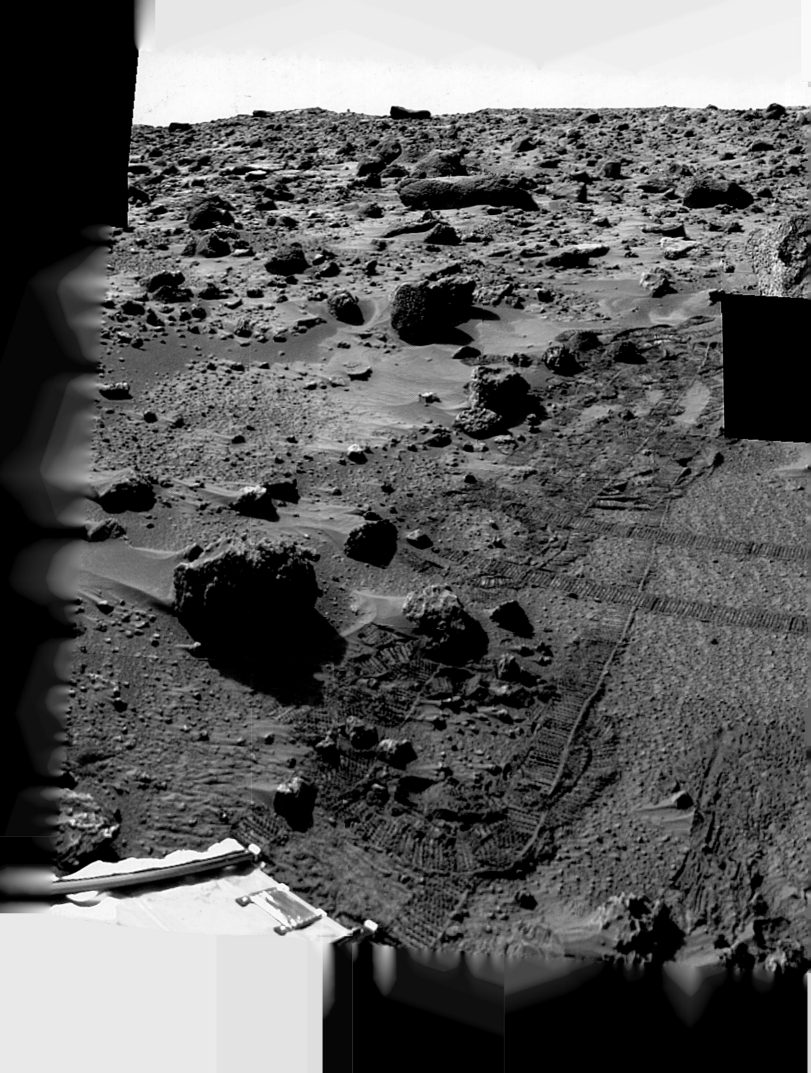

" I DON'T SEE ANY DATA "

S hortly before 2 a.m. on December 4, 1996, the night sky over Cape Canaveral, Florida, was illuminated by the roaring exhaust of a Delta II launch vehicle carrying an inter- planetary spacecraft unlike any other in the history of space flight. For the first time in more than 20 years, the United States was going back to the surface of Mars. One of Earth's closest neighbors, Mars, the fourth planet from the Sun, is a cold and desolate place, its surface a rust-red desert. Yet it is the one planet, aside from Earth, that could harbor life. Perched atop that Delta II rocket was a Mars-bound spacecraft called Pathfinder. It was not designed to look for life, but it was a new kind of spacecraft intended to inaugurate a Martian exploration program that could ultimately lead to a manned mission early in the 21st century. Pathfinder would give scientists their first look at the ruddy surface of Mars since 1976, when the twin Viking spacecraft returned the first pictures from the Martian surface.

Early portion of a super panorama sent to Earth by Pathfinder includes the end of the rear rover ramp, rover tracks, and the rocky surface of Mars. Barnacle Bill, the dark rock at left, exhibits a large wind tail of fine-grained atmospheric dust.

The Pathfinder launch was a success. After one delay because of bad weather and another to correct a minor software problem, Pathfinder was launched and began its 309-million mile journey to Mars. When the 1,973-pound spacecraft soared aloft, Mars was dimly visible on the horizon. Flight directors at Cape Canaveral turned the spacecraft over to mission controllers at the Jet Propulsion Laboratory (JPL) in Pasadena, California, and communication with Pathfinder was quickly established. A cheering mission control team prepared to begin monitoring data.

JPL engineers were just settling in when, unexpectedly, something went wrong. "As soon as we started getting data, that's when we figured out that we had this big problem," remembered Richard Cook, Pathfinder's mission manager. Cook asked each of the engineers monitoring the flight to report on the status of the mission. "You go around the room, and everybody says, 'Yeah, it's working fine.' Until one of them said, 'I don't see any data!' We started looking at the situation, and slowly this feeling builds up: My God, this is it. This is the end."

Before Pathfinder could do anything else, it had to check its position compared to the Sun and tell mission controllers which way it was pointing. A sensor that was supposed to check the position of the sun and report on Pathfinder's orientation was not working; it was not providing any data. The problem was serious. If Pathfinder failed, the consequences for NASA's Mars exploration program would be devastating. The agency was determined to prove that it could slash the costs of planetary missions while continuing to bring back high-quality science. Pathfinder was supposed to provide the proof. If Pathfinder failed, the Mars program could face years of delay. The failure of the sun sensor to provide data could doom not only the Pathfinder mission, but the entire NASA Mars program.

Pathfinder was a radical departure from the Viking mission, the previous mission to Mars, almost 21 years earlier. Indeed, Pathfinder was a departure from anything NASA had previously attempted. The Viking mission, one of NASA's most ambitious and sophis-

ticated interplanetary missions, cost one billion dollars in the 1970s. That's more than three billion dollars in 1997 dollars. Pathfinder's total budget was a scant 265 million dollars—less than one-tenth the cost of Viking. That included 55 million dollars to launch the spacecraft, 14 million dollars to run the mission, and, for the first time, a remote-controlled rover built at a cost of 25 million dollars. If the mission proved successful, JPL scientists would be able to drive around the Martian surface—by remote control—looking for the most interesting features. Pathfinder's mission designers had winked at Viking as they put together the mission: they planned to land the spacecraft on July 4, a date Viking 1 had tried for and missed.

Pathfinder was a three-foot-tall tetrahedron, with three triangular sides and a base. It was equipped with more computing power than any spacecraft in history, and it was crammed full of equipment. When it was loaded with fuel and ready for launch, the spacecraft, despite its compact size, weighed just under a ton (1,973 pounds). It included an imager, magnets for studying the properties of Martian soil, and wind socks and other instruments for studying the Martian atmosphere and weather. Its tiny rover measured two feet long by one-and-a-half feet wide and stood about a foot tall. It weighed a 23 pounds, but it was equipped with a camera and enough instrumentation to make a variety of geological observations.

Originally, the aim of the Pathfinder mission was simply to see whether it was possible to build a sophisticated spacecraft within a budget and on a schedule that were far more severe than had ever been imposed by NASA. Engineers at the Jet Propulsion Laboratory, the world leader in the design of planetary spacecraft, were energized by the challenge. A special team was assembled to build and operate Pathfinder and was given license to break all the rules—to re-invent spacecraft design and construction. Their efforts would become an example of American engineering knowhow at its finest. They amply demonstrated that JPL engineers had the right stuff.

In early December 1996 as Pathfinder was being readied for flight, its designers and builders waited anxiously for the launch and the confirmation that the spacecraft was on its way to Mars. In fact, just prior to the launch, engineers were called on to solve an unexpected problem that threatened to make one of Pathfinder's instruments unusable.

As the spacecraft was being folded into its launch configuration, its surface weather station was damaged; no one is quite sure how. A wire on a sensor designed to measure wind direction on the surface was accidentally severed. "Our hearts sank," said Tony Spear, the project manager for the Pathfinder mission. "We were emotionally crushed. We did not find the damage until just two hours before we were to fold the lander shut." Spear did not think there was time to fix the damage. It looked as though Pathfinder would have to be launched with the broken wind sensor. When word of the problem was relayed to the Pathfinder team at JPL, a member of the team, Regina A. Alleruzzo, asked for a chance to fix it. Alleruzzo, an expert in spacecraft wiring, was flown overnight to Cape Canaveral, where she was able to make the delicate repair under nearly impossible conditions.

Immediately after launch, when the problem with the sun sensor became apparent, the future of the mission seemed to be in jeopardy. Pathfinder's bright and brash young team of scientists and engineers began wondering whether their confidence and optimism had been misplaced.

The director of the Pathfinder mission, Tony Spear, had gone to Cape Canaveral for the launch, leaving Cook behind in Pasadena to run the mission. "I was catching the football," Cook said. "We were the people who controlled the spacecraft after launch." During the craft's ascent, Cook and his colleagues didn't hear anything from the spacecraft. That was according to plan: the rocket was sending information back to mission control at Cape Canaveral. Contact with the spacecraft was supposed to be made about 90 minutes after launch, when Pathfinder would separate from the upper stage of the rocket and turn on its transmitter. That's when NASA's Deep Space Network—the agency's network

of ground stations around the world—had to find the spacecraft and establish a communications link. "For me, that was the defining moment," said Cook, who had never run a mission before. "The problem is, you don't know where the spacecraft is. The launch vehicle is good, but it's not perfect, so you need to do a kind of blind search. And a blind search is a scary term. I could see this thing going out there, and we wouldn't be able to find it."

Operators at the Deep Space Network's Goldstone station in California were the first to report. They locked onto the rocket's third stage, and waited for the signal from the spacecraft. At mission control, Cook, wearing a headset, waited for a voice report from Goldstone. "We have X-band acquisition at DSS-15," Goldstone reported, raising cheers throughout the control room. Goldstone had locked onto Pathfinder's signal, which was being sent on a channel in the "X" frequency band. DSS-15 was the designation for the antenna at Goldstone that was picking up the signal. As the cheers subsided, Cook received the report from one of his engineers that the sun sensor aboard Pathfinder was not returning any data. Cook, who could call up anyone's data on his master computer screen, immediately switched screens to monitor the problem. "I thought it was a little bit strange," Cook said. He immediately directed mission controllers to try to determine the cause of the problem.

The sun sensor is a round disk, a little larger than a hockey puck, with a grating on the top and a series of photoelectric cells underneath. When sunlight falls on the sensor, the grating allows some of it to reach the photocells underneath. Depending upon the angle of the spacecraft, the sun will strike only some of the cells. As the spacecraft changes position, some photocells fall into shadow and others are newly exposed to the sunlight. The spacecraft uses that information to monitor and correct its orientation, spin, and trajectory during flight. Pathfinder's controllers quickly determined that not enough sunlight was reaching the sensor. Apparently, an explosive device used to separate Pathfinder from its rocket had deposited soot on the sensor, dimming the sunlight it received. The amount of light penetrating the sooty sensor was below what Pathfinder was programmed to recognize. Sensing a problem, Pathfinder ignored the readings from the sensor and failed to relay them to Earth.

"The only thing you really care about on your way to Mars is figuring out what your orientation is and changing your trajectory so that you can get into the right trajectory," Cook said. "But in order to change the trajectory, you have to know how you're pointed." Pathfinder was designed to spin slowly as it sped toward Mars, and the sun sensor would tell ground controllers how fast the spin was and whether the spacecraft had been correctly detached from the rocket.

After some discussion, mission engineers devised a solution. They would reprogram Pathfinder to lower the threshold at which it would conclude that the sensor was working. "We very quickly figured out how to get the software working, and we were excited," said Cook. "It was fixed, we were going to go on with the mission, and everything was wonderful." Cook and several others came in on a Saturday morning to transmit the files that would allow Pathfinder to accept the data from the sun sensor. He told his wife he would be home in a couple of hours.

"We had these three files we were going to send up, and we said, no problem, we'll just dash 'em off. So we send them up, and they don't get there. The spacecraft rejects them. So we send them again and again, and it just doesn't work." Very short files seemed to be reaching the spacecraft, but longer files were not being accepted.

As Cook was struggling with the files, the controllers monitoring spacecraft communications said they were noticing a dip in the signal from Pathfinder every five minutes. That was puzzling; there was nothing on the spacecraft or on the ground that was happening every five minutes. "We figured that was unrelated," Cook said.

As the problem continued, the population of the control room started to grow. As things continued to go wrong, Cook called Tony Spear in, and soon, as word of the trouble spread, others began to trickle in, including Donna Shirley, the manager of NASA's Mars Exploration Program. "Pretty soon, there are a zillion people here trying to figure out what's going wrong," Cook said. "I wouldn't say panic had set in, but it was beginning to."

The assembled multitude decided the files might reach the spacecraft if they were broken into tiny fragments. The three files that Cook was trying to send to Pathfinder were broken into about 30 tiny files. "Some of them get in and some don't, and so we start trying to change all kinds of things with the telecom configuration to make sure they're all doing the right things. By this time, about 12 hours had passed, and I called my wife and said, 'Yeah, I was supposed to be home about 9 hours ago, but I'm still having a problem.'"

One of the telecommunications specialists, meanwhile, had continued to think about the strange five-minute dip in the signal from the spacecraft. Perhaps it was related to the file-transmission problem after all. As he was trying to discover the source of the five-minute dip, the problem with Pathfinder appeared to worsen. It was not only files that were being rejected—Pathfinder was now starting to reject direct commands sent from mission control. "There's nothing worse than that, from our point of view, because our only ability to make something happen is to be able to send a command to the spacecraft. If it's not listening to us, that's it. If we lost communication, it was going to be end-of-mission."

The telecom specialist then realized that the five-minute dip was a misleading consequence of the way the signal from the spacecraft is measured. What was actually happening was that the signal was dipping momentarily with every spin of the spacecraft, which was spinning at about 12 revolutions per minute. Periodic measures of the spacecraft signal were catching the dip only once every five minutes, but it was happening about every five seconds. That was why files and even commands were not reaching the spacecraft—there was an interruption in the signal every five seconds. Files or commands that took longer than five seconds to transmit were not getting through. Shorter files would get through after multiple transmissions when, by chance, they happened to be sent inside one of the five-second windows. "There was some piece of metal or something else that was causing the signal to glitch every time the spacecraft went around," Cook said.

Once the problem became clear, Cook and his team had a solution. They decided, paradoxically, to send data to the spacecraft at a much slower rate. Even though commands and files would then take longer to reach the spacecraft, the slower transmission rate would allow Pathfinder's error-correction system to accept the transmissions. The interruption in the antenna signal was the result of something on the spacecraft that hadn't been taken into account when the antenna was designed, Cook said. The antenna had been tested before it was attached to the spacecraft, and it had worked fine. But it wasn't tested after it was attached to Pathfinder. "The antenna pattern can change if you have a piece of metal nearby; it doesn't have to be directly in the way of the antenna," Cook explained. "It changes the antenna pattern enough that it causes the glitch to occur."

The data rate was dropped from 250 bits per second to 7.8 bits per second, the slowest speed available. (Personal computers typically send data over telephone lines at 28,800 bits per second.) "The commands started getting in," Cook said. "It was a very tense day. We were there 20 hours—but it worked."

As soon as the files got in, the sun sensor began working and engineers on the ground could determine the spacecraft's orientation. Once that information was in hand, Cook and his team simply turned the spacecraft to face closer to Earth, and the antenna problem disappeared. Communications could resume at the regular rate.

Neither the sun-sensor problem nor the antenna problem alone would have been difficult to solve, but the two problems together were nearly fatal. It was what engineers call a multiple fault. "One of the things you talk about when you're designing spacecraft is single-point failures versus multiple faults,"

Cook said. Engineers ordinarily try to design spacecraft so they don't have a lot of single-point failures—individual problems that would each be enough by itself to doom the mission. But Pathfinder was different. "Pathfinder had all kinds of those," Cook said. That was the consequence, in part, of a new NASA mantra that had been established for spacecraft design—faster, better, cheaper. No longer would spacecraft be sent in pairs, as was done with the earlier Mariner and Viking flights to Mars, and no longer would money be lavished on redundant designs intended to prepare for every imaginable potential failure. Instead, Pathfinder was given enough flexibility and computer power so that it could detect and correct problems on its own. Pathfinder's designers tried to give the spacecraft flexibility to recover from a variety of single-point failures. "We said multiple faults would never happen," Cook said, chuckling. "And, right out of the bag, there goes a double fault."

Tony Spear later said that the communications problem, occurring so early in the mission, was vital to the success of the mission. Many of Pathfinder's designers and mission controllers had been children during the Viking flights to Mars. They needed a little testing. "Tony's view is that that's really where we became a team," Cook said. "We learned what it was really going to be like to do this mission, that it wasn't all going to be cake and ice cream—that it was going to be pretty tough."

Almost 21 years before Pathfinder was launched, the Viking mission had marked a triumphant end to the first phase of the exploration of Mars. Pathfinder was the first of a new generation of spacecraft that would inaugurate a new era. NASA was starting to talk about a so-called sample-return mission to bring Martian soil back to Earth. Even the notion of a manned mission to Mars was being discussed. If Pathfinder had failed, the future of Mars exploration could have been thrown into doubt.

Pathfinder went on to become one of NASA's most successful and spectacular planetary missions. The new age of Mars exploration had begun.

WHY DO WE DREAM OF MARS?

ars has held an unshakable grip on the human imagination for thousands of years. As ancient astronomers watched the stars sweep majestically across the black dome of the sky, they focused their attention on a strange reddish object that didn't seem to follow the laws of the heavens. Nearly all the stars seemed to move along the same arc, but Mars was one of only a handful of stars that did not follow the proper trajectory. The Greeks called these objects *planetes*, meaning "wanderers." Mars stood out from the other "wanderers" and not just because of its striking color. All of the visible planets—Mercury, Venus, Mars,

Because of the bloody color of a certain bright planet, the Romans named it Mars, after their god of war, often depicted as a powerful and terrifying character (opposite and above).

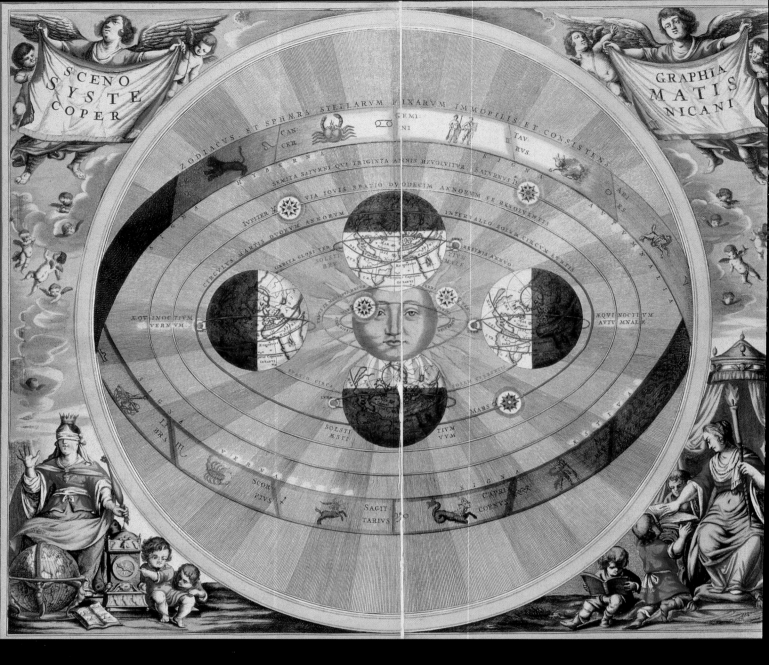

The Polish astronomer
Nicolaus Copernicus
(1473-1543) declared
in his masterwork,
*De Revolutionibus
Orbium Coelestium
(On the Revolutions of
the Celestial Spheres),*
that the sun, not the
Earth, was the center of
the universe. In drawings
and charts like the one
above—Scenographia:

Systematis Copernicani
Astrological Chart (ca
1543)—Copernicus
(left) showed that the
motions of the celestial
bodies more easily could
be understood if the
Earth revolved around
the sun, rather than the
sun around the Earth,
a notion that had been
the prevailing one since
ancient times.

About A.D. 140 Claudius Ptolemy (below), a scholar from Alexandria, Egypt, expanded Eudoxus's theory of the planets that put Earth at the center of the solar system. Although incorrect, Ptolmy's theory survived another 1,500 years.

Jupiter, and Saturn—drifted slowly in a narrow band across the constellations of the zodiac. With the exception of Mars, the visible planets all appeared to move in a constant direction across the constellations. The motion of Mars was more complicated, however. After moving in one direction for months, it would reverse course, moving backwards through the zodiac. This strange motion led the Egyptians to call it "the backward traveler." (Jupiter and Saturn do this too, but because they are farther away, they move more slowly and the backward motion is less noticeable.)

The strange backward motion was partly responsible—along with the planet's blood-red hue—for the early association of Mars with war. To the Greeks, the planet's wavering motion suggested distrust and disorder—with war as the ultimate expression of that disorder. The Greeks named the planet after Ares, the god of war. The Romans later renamed it after their god of war: Mars.

Little more was learned about Mars until the Renaissance, which marked the dawn of modern astronomy. During the Renaissance, Mars played an important role in one of the fiercest intellectual battles in the history of Western civilization: the debate over whether Earth was the center of the universe.

The Greeks were the first to place the Earth in that exalted position. As they watched the heavens revolve above them, they constructed a marvelous theory in which the Earth sat at the center of what they thought was a perfectly symmetrical universe. The mathematician and astronomer Eudoxus was one of the first to propose a theory of the universe. He advanced his ideas in the fourth century B.C. It was a simple, harmonious picture in which the planets moved in perfect circles around the Earth.

For a time, the theory put forth by Eudoxus seemed to work—it provided a simple explanation of what early astronomers saw in the sky above them. But, as often happens in science, the theory began to crumble under the weight of new observations. Astronomers making ever more careful observations of planetary motion began to find imperfections in the orbits of the planets. The planets had refused to move in perfect circles, as Eudoxus thought they must. The Greek theory of the solar system was faltering.

Rescuers soon appeared, however, to save the theory. The only way to truly explain the motions of the planets was to recognize that the Earth was not at the center of the universe. But that fact was such an alien notion that few of the ancients even considered it. Earth remained at the center of the universe, and new theories continued to explain what was happening around it. Aristotle struggled to modify the theory so it could explain the new observations. He made small revisions that helped the theory survive, but he did not resolve all of the outstanding questions. The theory found its greatest expression about A.D. 140, when it was revised by Claudius Ptolemy, a scholar in the Greek city of Alexandria, Egypt. Ptolemy discarded the circular orbits of Eudoxus and Aristotle. To describe the motion of the planets more accurately, he devised a complex system of curves he called "epicycles" and "eccentrics." Ptolemy drew charts and diagrams showing how these curves could be used to describe the motions of the planets more accurately than the simple circular orbits of Eudoxus did.

It was a master stroke. Ptolemy's theory of the solar system, with its epicycles and eccentrics, was brilliantly persuasive, surviving mostly intact until the 16th century. That was 2,000 years, more or less, after Eudoxus proposed it. The theory was embraced by, among others, the Roman Catholic Church, which found evidence for an Earth-centered universe in the Bible. To argue that the Earth was not the center of the

Johannes Kepler (below) discovered laws of planetary motion that finally put to rest the tenacious idea that the sun and the entire universe revolved around the Earth. Kepler's breakthroughs in understanding and testing the principles underlying those planetary laws laid the foundation for modern physics.

universe was heresy, and those who did so risked the death penalty.

Ptolemy's theory had one unfortunate drawback: It was wrong. Even before Ptolemy succeeded in propping up the crumbling theory of Eudoxus and Aristotle, one lone voice was raised to point out the error. In the third century B.C., Aristarchus of Samos, a Greek astronomer, suggested that the planets moved around the Sun, not the Earth. His name survives because he was mentioned in one of the works of the great Greek mathematician Archimedes. Aristarchus—whose own writings are lost—watched the shadow of the Earth as it swept across the moon during a lunar eclipse. He determined that the sun must be much larger than the Earth and that it appeared small only because it was far away. He thought it unlikely that the giant sun would revolve around the much smaller Earth. Therefore, he concluded, the sun must be at the center of the solar system. And the Earth and the other planets must be moving around the sun.

Aristarchus was condemned for this heretical notion. As the years passed, and Ptolemy's theory remained universally accepted, students of the classics continued to encounter Aristarchus. But few accepted his theory of the planets. Or, if they did, they didn't risk saying so, given the Catholic Church's position. As a result, the work of Aristarchus did little to dampen the enthusiasm of most of the Ptolemaic theory's proponents.

In the 16th century, however, the Polish physician and lawyer Nicolaus Copernicus published the results of groundbreaking astronomical research he had conducted in his spare time. He made very careful observations of the motions of the planets, and he struggled to make sense of them. In his book

De Revolutionibus Orbium Coelestium (On the Revolutions of the Celestial Spheres), he announced the results of his deliberations and calculations. The Earth, he concluded, could not possibly be the center of the universe. A careful analysis of his observations led Copernicus to what seemed an inescapable conclusion: that the sun must indeed be at the center of the solar system, with the Earth whirling in orbit around it. Copernicus, who had worked as a church administrator, dedicated his book to Pope Paul III.

The book was published in 1543, when the author was on his deathbed.

Copernicus's radical proposal was a major advance in astronomy, but it did not solve all of the problems Copernicus encountered in explaining the planets' movements. Copernicus had held on to the ancient idea that the planets moved in circles. Putting the sun at the center of the solar system had solved many of the problems of earlier theories, but not all of them. Copernicus had begun the job of dethroning Ptolemy, but he hadn't quite finished the work. And this is what brings us back to Mars. For all the strengths of the Copernican system, it failed to explain the motion of Mars, including its occasional, peculiar backwards march across the constellations of the zodiac.

Some 60 years later, the failures of the Copernican system attracted the attention of the Danish astronomer Tycho Brahe. Born to nobility in Denmark three years after the death of Copernicus, Tycho established the finest astronomical observatory in Europe. In 1600, he invited the German mathematician Johannes Kepler to join him in Prague, where he put Kepler to work calculating a new orbit for Mars. The red planet was chosen because its movements were among the most difficult to explain.

QVADRANS MVRALIS
SIVE TICHONICVS.

Tycho Brahe established Europe's finest observatory on the island of Hven in the late 16th century. In this 17th century illustration, he demonstrates the wall quadrant, one of many astronomical instruments he invented. With the wall quadrant Tycho Brahe could take precise measurements of a celestial object's position. Such observations revolutionized astronomy and laid the foundations for today's laws of planetary motion.

Early astronomers found wonders in the heavens
that they described with great accuracy and
recorded in wondrous detail (above). But because
they lacked a basic understanding of the motion of
celestial bodies, they had difficulty explaining what
they saw. The intense curiosity that motivated them

Perhaps the greatest of the early astronomers, Galileo (below), like Kepler, engaged astronomical challenges as a theoretician, aiming not only to observe the heavens but also to explain them.

Kepler labored over a mathematical description of the orbit of Mars for years. He imagined himself on Mars and tried to calculate the motion of Earth from that vantage point. He then tried to calculate the motion of Mars as seen from the sun—and that's when he made his breakthrough. In 1609, he triumphantly announced that the orbit of Mars was not a circle, but an ellipse. In his historic book *Astronomia Nova*, he also explained his first two laws of motion: that all planets move in elliptical orbits with the Sun at one of the two foci of the ellipse, and that orbiting planets sweep out equal areas in equal times. (Or, in other words, planets move faster as they get nearer the Sun and more slowly as they move farther away.)

Observations of Mars in the 17th century helped to solidify the triumph of the Copernican system. Some of the key findings were made by Galileo, who noted that the diameter of Mars as seen through the telescope seemed to vary, reflecting differences in the distance between Mars and Earth. The same was true for Venus. Galileo used those observations and others to argue forcefully in favor of the Copernican system. "The problem for those who knew their Ptolemy was that Ptolemy's planetary theories of Mars and Venus predicted those differences as well," says Albert Van Helden, a professor of history at Rice University and the originator of a World Wide Web site called The Galileo Project, a storehouse of information on the great astronomer. Nevertheless, Galileo succeeded in finally demolishing the Ptolemaic system, after a wrangle with the Church, which put him on trial in 1633 for his heretical views. During the trial, Galileo publicly disavowed his conclusions, in what amounted to a plea bargain. Nevertheless, his ideas triumphed. Kepler's laws of motion helped set the

In 1877, the Italian astronomer Giovanni Schiaparelli (below) reported that he had discovered a strange network of lines on the surface of Mars. He called these lines *canali,* an Italian word meaning "channels" or "canals." Schiaparelli did not suggest that intelligent beings built the canals, but others quickly did so. And many people gained a strong conviction that Mars was home to intelligent life.

stage for the monumental *Principia* of Sir Isaac Newton in 1687, perhaps the most important scientific work in history. The intellectual warfare over Earth's position in the universe was over, and the planet of war had played an important role. Today, the notion that the Earth is at the center of the universe persists only in the archaic language we use to describe the motion of the sun. As the late planetary scientist Carl Sagan of Cornell University pointed out, we still talk of the sun rising and setting. "It is 2,200 years since Aristarchus, and our language still pretends that the Earth does not turn," he said.

While the battle over the Ptolemaic and Copernican world systems was raging, Kepler took time away from his scientific work to write a book called *Somnium (The Dream).* Published in 1634, the book was an outgrowth of an essay Kepler had written to introduce the Copernican theory to a wider audience. In *Somnium,* one of the first works of science fiction, Kepler imagined a journey to the moon. He described its inhabitants and explained how they survived in the moon's extremes of hot and cold, and he described the Earth as it would appear to these mysterious lunar dwellers. Later in the 17th century, other writers began to envision space journeys and life on other worlds, and generations of science fiction writers have followed. Some, like Kepler, have taken their readers to the moon. But Mars soon became the favored destination, especially at the end of the 19th century, when increasingly powerful telescopes convinced astronomers that the moon was barren. Mars was far more difficult to see with the best telescopes of the day. Astronomers were unable to determine much of anything about the surface of the ruddy smudge that appeared in telescope eyepieces. Many unanswered questions about the planet persisted, offering science-fiction writers an irresistible opportunity to invent answers. Many writers found themselves inescapably drawn to Mars, still as alluring and mysterious as it had been to ancient observers.

Speculative stories about Mars have made the planet seem so real that fiction and fact have, at times, become hopelessly entangled. This question was addressed by an unusual gathering of scientists and science-fiction writers at the California Institute of Technology in Pasadena on November 12, 1971, the evening before the U.S. spacecraft Mariner 9 was due to go into orbit around Mars. The participants were the noted planetary scientists Bruce Murray of Caltech and Carl Sagan of Cornell University, the science-fiction writers Ray Bradbury and Arthur C. Clarke, and Walter Sullivan, then the science editor of the *New York Times.* "Mars somehow has extended and endured beyond the realm of science to so grab hold of man's emotions and thoughts that it has actually distorted scientific opinion about it," said Murray. "So it isn't just the popular mind that has been misled, but the scientific mind as well.... We *want* Mars to be like the Earth. There is a very deep-seated desire to find another place where we can make another start."

The famous Martian "canals," once believed to be evidence of a highly evolved society, turned out to be a beguiling fiction produced not by science fiction writers but by scientists themselves. The apparent network of lines on the Martian surface was first reported in detail by the Italian astronomer Giovanni Schiaparelli in 1877. He called them *canali*—a word that can be translated to mean

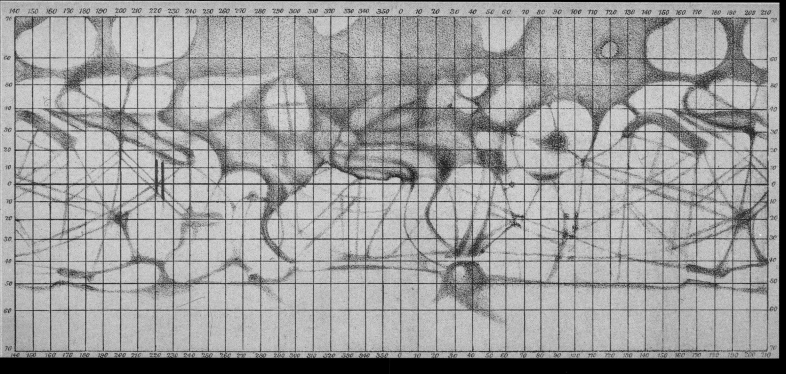

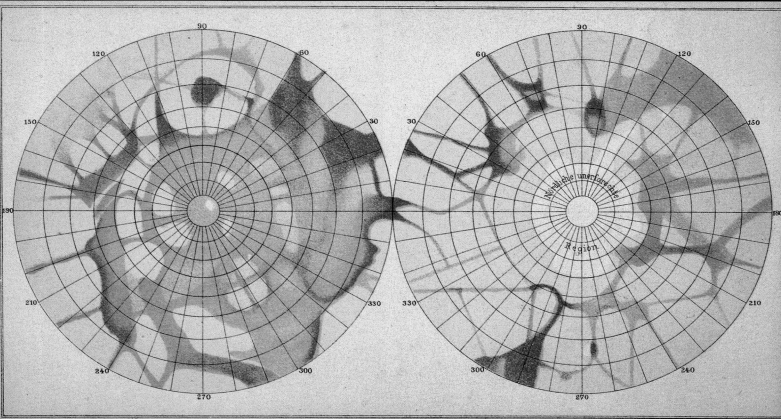

Schiaparelli's maps and drawings of the features he thought he saw on the Martian surface show a confusing mosaic of unexplained characteristics. Telescopes in use in the late 19th century provided unclear images that defied easy analysis. Much of the interpretation relied on the imagination of the astronomers who peered through those early instruments. Some thought the crisscrossing lines shown near the middle of Schiaparelli's map (upper) were too long and straight to be accidents of nature. Yet other astronomers could not even find these features, and many dismissed them.

"canals" or simply "channels." The discovery excited the enthusiasm of the American scholar and self-taught astronomer Percival Lowell, a member of the distinguished Lowell family of Boston. Lowell was born in 1855, but he did not take up astronomy seriously until the 1890s, when he was almost 40 years old. He quickly dismissed the subtleties of translating *canali*, and from that time on they became canals. Lowell turned his prodigious intellect and inexhaustible energy to studying the canals and stoking popular interest in them, and he made them into the greatest subject of wonder in the heavens. Although the canals were ultimately dismissed as the product of Lowell's misguided scientific imagination, they helped make the search for life on Mars a central concern for scientists as well as for writers of science fiction.

The Pathfinder mission of 1997 was the latest in a series of exploratory voyages that were intended to conduct basic scientific studies of Mars. Many of them have also been shaped largely by the question of whether life exists there. As scientists continue to collect the data to answer that question, writers, not bound by the bothersome constraints of scientific research, have marched boldly into the void, finding life of all kinds all over the red planet.

Among the first writers to craft a novel of Mars that has endured was H.G. Wells, whose *The War of the Worlds*, published in 1898, portrayed Martians as bloodless, technologically advanced invaders, who exterminated thousands in an attempt to take control of Earth. A century after it was published, the novel is still widely read. It was also the inspiration for the most famous radio play in American history. The 1938 broadcast of *The War of the Worlds* by Orson Welles incited panic across the country, as millions of listeners mistakenly took Welles's fictional news reports of a Martian invasion to be real.

Wells was followed by Edgar Rice Burroughs, better known as the creator of Tarzan. Burroughs's *A Princess of Mars*, published in 1917, was the first of a series of 11 novels about Mars. They recount the adventures of John Carter and his family and friends,

This mosaic of the eastern hemisphere of Mars centers on the crater named for Schiaparelli, the 19th-century Italian astronomer who intensely studied the Martian surface. Almost 250 miles (400 km) across, the crater is located 340°W, 2°S. Ancient, heavily cratered terrain dominates this view, which also shows a dark wedge-shaped region stradling the equator. The dark areas are probably devoid of the bright red dust that gives Mars its distinctive color. Globes such as this are produced by merging high- and medium-resolution black-and-white Viking images with low-resolution color images and placing them in a global projection.

Edgar Rice Burroughs (above) wrote 11 novels about Mars, but most readers know him best for his tales of Tarzan. His books on Mars, a planet that his fictional Martians called Barsoom, inspired the science fiction writer Ray Bradbury and the astronomer Carl Sagan.

who travel to Mars to wage battle for the forces of good. At the 1971 symposium at Caltech, Carl Sagan and Ray Bradbury both confessed to being inspired by Burroughs. "I myself first became aware that Mars was a place of some interest by reading stories by Edgar Rice Burroughs," Sagan said. He recalled trying to imitate John Carter, who traveled to Mars "by standing in an open field and sort of spreading his arms out and wishing." Sagan, at age 8 or 9, failed to reach Mars, "not entirely to my surprise," he said. "But I thought there was always a chance." That boyish optimism about space remained with Sagan throughout his career, which ended with his death in 1996.

Bradbury compared Burroughs to Jules Verne, whom he called a moralist, the author of "adventurous moralities." Burroughs, on the other hand, was "urging boys to take off their skins and dance around in their bones. He was always cutting people's heads off and leaving the bodies there." Burroughs "was first and foremost the vulgarian who took me out under

the stars in Illinois and pointed up and said, with John Carter, simply: Go There." Bradbury went on in 1950 to write *The Martian Chronicles*, a science fiction classic considered by many to be the best book on Mars ever written. It captured the romance of Mars as well as anything before or since, and it inspired, among others, Arthur C. Clarke, best known for *2001: A Space Odyssey*. In 1952, Clarke wrote *The Sands of Mars*, one of the first novels to talk about transforming the cold, dry, Martian surface into a habitat suitable for human life—a process that has come to be known as terraforming. Mars continues to inspire countless works of fiction that all trace their origins back through Bradbury to Burroughs. As Bradbury has remarked, "Without Edgar Rice Burroughs, *The Martian Chronicles* would never have been born."

These books and the hundreds of others written about Mars vary widely in their depiction of Martian life and Earthlings' response to it, but they share several qualities. The best of them use Mars as

Martian spacecraft attack Earth in *The War of the Worlds,* a novel of suspense by H. G. Wells. The Martian invaders shower terror on Victorian England with weapons of mass destruction. Chance, not human resourcefulness, ultimately saves the planet.

Wells, Welles and The War of the Worlds

On October 30, 1938, a Sunday afternoon, 23-year-old Orson Welles met with the cast of the weekly CBS radio show *Mercury Theater on the Air*. At that meeting, Welles got his first look at a script written by Howard Koch, a would-be screenwriter who went on to become co-author of *Casablanca*. Welles and his partner, John Houseman, had suggested that an adaptation of the science fiction novel *The War of the Worlds*, by H.G. Wells, would make a suitably frightening show for the night before Halloween. Asked to produce the script in less than a week, Koch had endured sleepless nights and endless rewrites as he retold the story of the Martian invasion of Earth. Welles read the script and prepared to go on the air with the show that Sunday evening. A modestly successful show, *Mercury Theater* was hurt by competition from the Edgar Bergen-Charlie McCarthy show on NBC radio; both aired Sunday at 8 p.m. eastern time.

The radio play began, as usual, with Welles as the narrator. He opened the program with what his biographer David Thomson calls that "voice of magnificent appropriation, the deep, smooth, cultured sound of…chocolate pudding.…" And this is what he said:

We know now that in the early years of the twentieth century this world was being watched closely by intelligences greater than man's and yet as mortal as his own. We know now that as human beings busied themselves about their various concerns they were scrutinized and studied, perhaps almost as narrowly as a man with a microscope might scrutinize the transient creatures that swarm and multiply in a drop of water. With infinite complacence, people went to and fro over the earth about their little affairs, serene in the assurance of their dominion over this small spinning fragment of solar driftwood which by chance or design man has inherited out of the dark mystery of Time and Space. Yet across an immense ethereal gulf, minds that are to our minds as ours are to the beasts in the jungle, intellects vast, cool and unsympathetic, regarded this earth with envious eyes and slowly and surely drew their plans against us. In the thirty-eighth year of the twentieth century came the great disillusionment.

Koch set the story not in England, where the novel took place, but in Grovers Mill, New Jersey, a small town near Princeton. The opening of the show made clear that this was a radio drama, but many listeners missed the opening, because they were tuned into Edgar Bergen and Charlie McCarthy on NBC. About ten minutes into that show, Bergen presented a singer who was not especially popular, and many listeners left Bergen for *Mercury Theater*. Welles's audience doubled. What the newcomers heard was a show being interrupted by news bulletins describing cataclysmic earthquake-like shocks in southern New Jersey. The information in the bulletins was sketchy at first; details emerged gradually with each subsequent interruption. That gave the bulletins a verisimilitude that listeners found easy to accept at a jittery time when Europe was moving quickly toward war and such news flashes were common. One of the actors reading the bulletins had taken his cue from a recording of a radio report of the destruction of the *Hindenburg* airship a year earlier.

H. G. Wells wrote many works of science fiction, but *The War of the Worlds* made his name a household word for all time.

As the dramatized bulletins continued, it became clear that the shocks marked the arrival of Martian ships, and that the Martians were attacking with an unstoppable, murderous ray. Soon New York City was reported to be in flames as the Martian march moved northward from Grovers Mill.

The New York Times received 875 calls. People were said to be abandoning their homes in panic, fleeing what they thought was the imminent arrival of interplanetary monsters. The radio drama ended with the Martians

> *"We know now that in the early years of the twentieth century, this world was being watched closely by intelligences greater than man's."* – **ORSON WELLES**

succumbing to bacteria to which they had no resistance. Welles ended the program by wishing listeners a happy Halloween, but by then it was too late. Those who believed the program was real were no longer listening. CBS, worried that it would face some legal liability, broadcast disclaimers all evening. The next day, Welles held a press conference at which he denied there was any deliberate attempt to cause a sensation, and there is no reason to doubt him. The broadcast demonstrated how willing Americans were in 1938 to accept the idea of intelligent life on Mars. They had grown up hearing stories about canals on Mars; the idea that the canal-builders were malevolent creatures capable of traveling to Earth was evidently easy for many to accept. The alarm quickly died down, and the only significant consequence of the broadcast was that it made a star of Orson Welles. *The War of the Worlds* helped Welles get to Hollywood, where, three years later, he made what is probably the most acclaimed American film—*Citizen Kane*.

Orson Welles (above) performs the radio adaptation of *The War of the Worlds*, whose location had been moved from England to New Jersey. The program triggered panic in listeners who tuned in midway through what seemed to be a news account of a Martian attack. The poster (below) dates from the 1953 movie, made by Paramount Pictures.

a stepping-off point to highlight social problems of the times during which they were written. *The War of the Worlds*, in addition to being a well-written, suspenseful adventure novel with a highly satisfying plot twist at the end, is a powerful satire on Victorian society. The ease with which the Martian invaders overran and destroyed England satirized the determination and brutal disregard for local people that often characterized the actions of English colonialists. *The Martian Chronicles* is a product of the Cold War, in which Martian society faces the same cataclysmic threat that the arrival of atomic weapons posed to Earth's survival. More recently, the novelist Kim Stanley Robinson's series of three novels— *Red Mars, Green Mars*, and *Blue Mars*—shows the futility of trying to create a Martian society that avoids the mistakes made on Earth. Robinson's novels explore the bitter division between the "reds,"

the Martian environmentalists who want to leave the planet mostly undisturbed, and the "greens," the developers and profiteers who want to terraform Mars, rebuild it, and exploit it.

These writers, and many more like them, share credit with scientists for keeping the romance of Mars alive. Thousands of years after the ancients gazed at Mars with fear and fascination, it remains among the most mysterious and intriguing bodies in the heavens. The exploration of the Earth's surface is nearly complete, and Mars represents a new frontier, a dazzling new destination to fire the imaginations of today's explorers. The technology to send a human mission to Mars is now within reach. And it is conceivable that, some time in the next century, scientists may join science fiction writers in the exploration of Mars, as human beings, for the first time, step onto the cold red dust of the Martian surface.

**This cartoonish, bubble-headed Martian hails from the 1996 Warner Brothers movie *Mars Attacks!*
With this depiction of Martians as impish killers, director Tim Burton spoofed the low-budget science fiction films
of the 1950s, known for their predictable plots.**

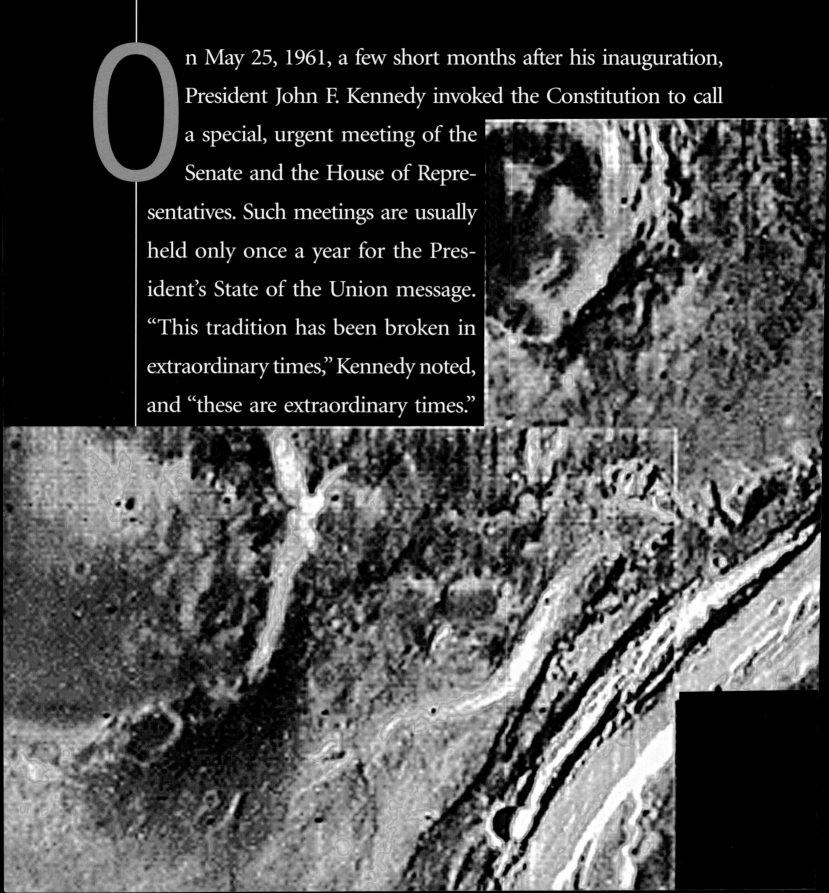

On May 25, 1961, a few short months after his inauguration, President John F. Kennedy invoked the Constitution to call a special, urgent meeting of the Senate and the House of Representatives. Such meetings are usually held only once a year for the President's State of the Union message. "This tradition has been broken in extraordinary times," Kennedy noted, and "these are extraordinary times."

Kennedy called for stepped-up efforts in the worldwide struggle against the Soviet Union and the Red Chinese. He called for economic and military aid to countries threatened by Communist takeover, and for measures to strengthen America's economy. But the speech is best remembered for a message Kennedy reserved until the end.

It was time, "for a great new American enterprise—time for this nation to take a

Three-image mosaic taken by the Mariner 9 spacecraft of part of the Mangala Vallis channels on Mars surprised scientists by showing a surface with an active and protracted geologic history.

The Mariner 4 spacecraft (opposite), launched on November 28, 1964, gave scientists their first close look at perhaps the most fascinating planet in the solar system. That look disclosed Mars as barren and crater pocked, with a landscape more like the moon's than the Earth's. Prospects for finding life on Mars seemed bleak. Initially, the Soviet Union's launch of the Sputnik satellite in 1957 (below) had spurred U.S. exploration of space.

clearly leading role in space achievement," he said. "I believe this nation should commit itself to achieving the goal, before this decade is out, of landing a man on the moon and returning him safely to Earth," Kennedy declared. And the race to the moon was on.

While those words were still echoing through the halls of Congress, Kennedy went on to call for additional funds to build a rocket "for even more exciting and ambitious exploration of space, perhaps beyond the moon, perhaps to the very end of the solar system itself." While the public was dreaming of the race to the moon, scientists took note of Kennedy's call for exploration of the solar system. "President Kennedy's Apollo man-to-the-moon initiative in May 1961 implicitly provided NASA (the National Aeronautics and Space Administration) carte blanche to race to the planets with robots as well," noted Bruce Murray, a Caltech geologist and former director of the Jet Propulsion Laboratory in Pasadena, California, headquarters of America's planetary exploration program. And NASA was ready.

Kennedy's speech was, of course, partly a reaction to the Soviet Union's early lead in space exploration. On October 4, 1957, the Soviets inaugurated the space age with the launch of Sputnik I, a basketball-size, 183-pound artificial satellite. It was the first object to orbit the Earth. The White House had announced in 1955 that it planned to lob a tiny, three-and-a-half pound payload into orbit during the International Geophysical Year, an international scientific research program that ran from July 1, 1957, to December 31, 1958. The Soviets easily beat that deadline, and they did so with a much larger payload. It was a dazzling engineering achievement, and

it was a powerful Cold War victory, raising fears in the United States that the Soviets might have the capability to strike the United States with a nuclear missile launched from Europe.

As NASA raced to design a program to put a man on the moon, it also began to plan exploratory unmanned missions to Mars. Kennedy's speech came nearly a century after Giovanni Schiaparelli had first reported finding "canals" on Mars. But in the intervening years, researchers had failed to make any significant headway in resolving the question of whether life existed there. In 1922 and 1924, as Mars passed close to Earth, the U.S. government asked radio stations to stop broadcasting long enough for radio operators to listen for signals from Mars. "Navy stations across the Pacific stopped transmitting on the night of August 22, 1924, and remained silent for three days, hoping to pick up a signal," observed science writer John Noble Wilford. William F. Friedman, the master codebreaker who would later lead the team that deciphered Japanese diplomatic code just before the attack on Pearl Harbor, was called in to interpret any messages that might be received. As Wilford noted, perhaps the most interesting thing about this episode is that the world seemed safe enough for the military to suspend communications for three days.

At the time of Kennedy's speech, astronomers had what they thought was a reasonably complete picture of Mars. They had determined that Mars had far less water than Earth and was without oceans. Water was believed to exist at the poles, where astronomers watched white, icy polar caps shrink and grow with the seasons. An astronomer named Earl C. Slipher, working at Lowell's Flagstaff, Arizona, observatory, had amassed more than 100,000 photographs of Mars. To the end, Slipher, who died in 1964, remained a firm believer in Lowell's Mars. "Since the theory of life on the planet was first enunciated some fifty years ago, every new fact discovered has been found to be accordant with it....Every year adds to the number of those who have seen the evidence for themselves," he wrote. Had Slipher lived but a few more years, he would have found his faith in Martian life sorely tested. By the time the space age began, most astronomers had long since abandoned the notion of canals on Mars, even if they were less sure about the existence of some form of life. Astronomers found it impossible to confirm the existence of the canals; some thought they saw them, and some didn't. Most astronomers eventually concluded that the canals were some sort of optical illusion, if they existed at all. Although Slipher continued to believe in the existence of the canals, he did make important contributions to the study of Mars. He left behind a superb map of the Martian surface, produced from his huge collection of photographs. The last edition of Slipher's map still included Lowell's network of canals.

One of the critical issues in Mars science in the early 1960s was whether the density of the Martian atmosphere could be accurately determined. In other words, was there enough atmosphere to support life, or was the air so thin that it would almost rule out the possibility of life? Lowell had tried to use measurements of the reflectance of light from the Martian surface to try to determine atmospheric pressure on Mars. He calculated that the pressure at the surface must be about 87 millibars, or roughly one-tenth that of the atmospheric pressure at Earth's surface (1,000 millibars). Other calculations supported these figures, and many astronomers believed Lowell was correct on this point.

Another critical issue concerned the makeup of the atmosphere. In 1947, Gerard P. Kuiper, a Dutch-American astronomer, identified the presence of carbon dioxide in the Martian atmosphere. Kuiper used a spectrometer to analyze the different frequencies of infrared light reflected from the surface of Mars. Carbon dioxide absorbs certain frequencies of infrared light, while allowing others to pass—leaving a telltale signature. Kuiper measured the infrared light coming from the moon, which has no atmosphere, and compared it with the reflected light from Mars. Once he had taken account of the absorption by gases in the Earth's atmosphere, he was left with measurements that showed that carbon dioxide existed in the Martian atmosphere. When he tried to calculate the atmospheric pressure of carbon dioxide on Mars, however, he made an error. His estimate was low, and it led him wrongly to conclude that the Martian polar caps could not be made of frozen carbon dioxide.

Later work appeared to confirm Kuiper's erroneous determination of the pressure of carbon dioxide on Mars, and it seemed increasingly clear that the polar caps must be made of water. In the mid-1950s, erroneous determinations of the overall atmospheric pressure came close to confirming Lowell's original measurements decades earlier. And Mars seemed to have weather patterns not entirely unlike those on Earth; clouds drifted through the Martian sky and huge, powerful dust storms kicked up on the surface. In 1956, astronomers watched a dust storm begin in the southern hemisphere and grow until it reached around the entire planet.

Astronomers snapping pictures of the Martian surface noted that many of the surface details were difficult to see when photographed through blue, violet, and ultraviolet filters. Some theorized the existence of a "violet layer" in the atmosphere that might shield the surface from deadly ultraviolet radiation. No one could quite imagine what such a thing might be made of, yet it persisted as a weird possibility until the exploration of space began.

In other respects, however, Mars was considered a twin of Earth. The tilt of its axis is nearly the same as the Earth's—and different from the other planets. That means Mars has seasons like Earth's. Its north and south polar caps shrink in summer and grow in winter, just as would be expected. Although some had speculated that the polar caps were made of

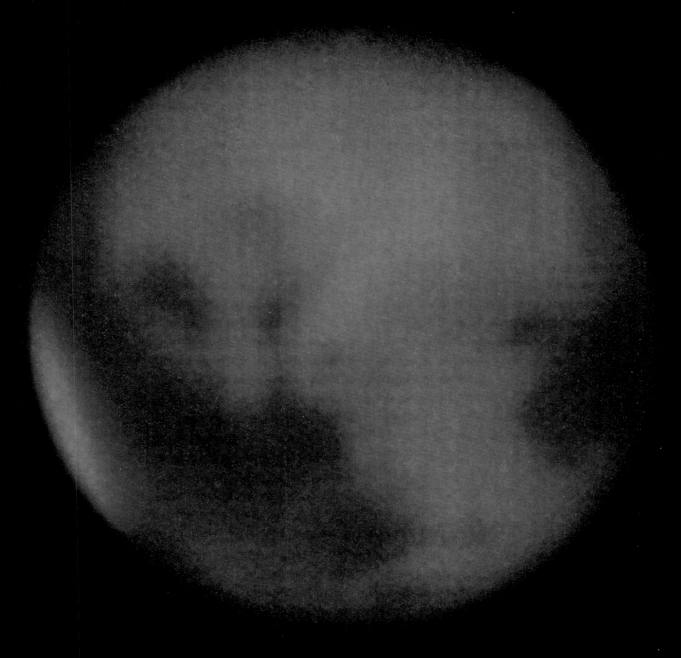

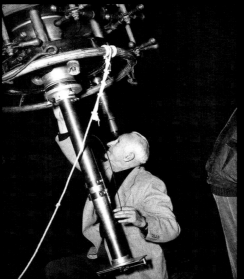

Earl C. Slipher (right), at work in the Lamont-Hussey Observatory in South Africa, amassed a total of more than 100,000 photographs of Mars (including the one above). Slipher died in 1964, just before Mariner 4 became the first spacecraft to obtain detailed images of Mars. In the late 1950s, when the U.S. Air Force prepared to send spacecraft to the red planet, it adopted Slipher's map of Mars because it contained the most detailed available information on that planet. Yet, to the end, Slipher remained a believer in the work of Percival Lowell, and the map used by the Air Force included Lowell's canals.

In 1954 astronomer Earl C. Slipher made an expedition, sponsored in part by the National Geographic, to this observatory in Bloemfontein, South Africa, for the best view of that year's close pass of Mars.

frozen carbon dioxide, the prevailing view then was that they were made of ice. And yet another similarity is that a day on Mars lasts about 24 hours, 37 minutes—nearly the same length as Earth's.

The most remarkable similarity between the two planets had to do with a pattern of light and dark regions, visible from Earth, that seemed to vary with the seasons. Despite the lack of liquid water on Mars, it seemed possible that this "wave of darkening," as it was sometimes ominously called, was produced by vegetation that flourished in summer and became dormant in winter, as much of Earth's vegetation does. Mysteriously, the dark regions were said to have a bluish or greenish cast. In 1954, Slipher, still pursuing Lowell's vision of life on Mars, led a National Geographic Society-sponsored expedition to Bloemfontein, South Africa, for a prime view of the planet, which would be in the southern hemisphere during a close pass by Earth. Slipher reported the discovery of a nearly Texas-size dark area that "appeared to have the same blue-green tint as previously known dark areas." Biologists speculated that the dark areas were caused by the growth of lichens, or something similar. "Such green areas bear eloquent testimony to the fact that Mars is not a dead world," said the indefatigable astronomer. "If this were not so, the winds of Mars would long ago have scattered the dust and sands everywhere, rendering the whole surface the same uniform tint."

One further piece of evidence—which also proved, ultimately, to be wrong—provided what seemed to be confirmation of Slipher's speculation. William Sinton, an astronomer who began work at Percival Lowell's observatory in Flagstaff and continued with the giant 200-inch telescope at Palomar Observatory near San Diego, analyzed the sunlight reflected from Mars. In 1958, he found that certain frequencies of light were being absorbed by something on the surface. These frequencies were characteristic of molecules containing bonds between carbon and hydrogen—or, in other words, organic molecules like those that make up living organisms on Earth. Tests were done on lichens and mosses, and, incredibly, they showed precisely the same absorption pattern seen on Mars. The areas where light was absorbed became known as the Sinton bands. "On the face of it, a stronger confirmation of Lowell and Kuiper could hardly have been imagined," said Norman Horowitz of the California Institute of Technology, who was the chief of biological sciences for the Mariner and the later Viking missions. The Space Science Board, an august group formed to advise NASA, did not entirely accept Sinton's findings. The group argued that Sinton had not done enough to rule out the possibility that the Sinton bands were caused by some combination of inorganic materials. Nevertheless, their conclusion would have cheered Lowell: "The evidence taken as a whole is suggestive of life on Mars."

NASA hoped to send its first missions to Mars using a version of the Saturn rocket that would later take astronauts to the moon. Budget cutbacks and development problems twice forced NASA to move to smaller rockets, however, and the Martian exploration program was delayed. The first planned spacecraft, at one point expected to weigh more than a ton, would have to be trimmed until it weighed no more than 575 pounds. Even at that reduced size, it would be required to carry a television camera and the communications gear needed to get pictures back to Earth, communicating across a distance far greater than had ever been attempted. By February, 1963—nearly two years after Kennedy's speech—NASA was trying to see whether it could build a Mars-bound spacecraft small enough to fit on an Atlas/Agena class rocket.

The Soviet Union, locked into the race to the planets, launched its first two planetary spacecraft in 1960; both failed. (As was their custom during the Cold War, the Soviets kept these launches secret, saving themselves the indignity of acknowledging failure or trying to explain what went wrong.) The Soviet Union launched its first Mars probe in 1962, putting the Soviets well ahead of NASA and apparently on track to arrive at Mars before their American counterparts left the ground. Halfway to Mars, however, the Soviet probe, named Mars 1, disappeared. Its signal was lost, and its fate remains unknown. The loss of Mars 1 was a distinct stroke of luck for Cold War America; it meant that NASA's hopes of being first to Mars were still alive.

NASA's plan for planetary exploration was to launch spacecraft in pairs, increasing the likelihood of reaching its goals. At the time, the Soviet Union's ability to launch vehicles into space surpassed that of the United States. NASA was feeling the pressure of time, and it faced a staggering engineering challenge. But as Bruce Murray has pointed out, the pressure on NASA had two fortuitous consequences. The Soviets had equipped their spacecraft with bulky chambers that maintained Earthlike pressure and temperature conditions to protect spacecraft electronics and other sensitive components. Unable to loft that much weight into space, NASA was forced to design its spacecraft to operate in the vacuum of space, with its alternately blistering and frigid temperatures. Secondly, emphasis on reliability was an absolute priority, forcing rigorous ground testing of everything that went into the American space probes. In the long run, these compromises, which were forced upon NASA, would prove crucial to the success of the American space program.

The first pair of American planetary explorers, Mariner 1 and 2, were bound for Venus, and were launched in 1962. Mariner 1 went tumbling into the Atlantic, the victim of a launch failure. Mariner 2 reached Venus, becoming the first spacecraft to visit another planet. It swept by the planet in December 1962, monitoring temperatures and taking other measures of Venus and its thick, choking atmosphere. Two years later, NASA was ready for its first voyage to Mars. Mariner 3 and Mariner 4 were prepared for launch in November 1964 when the memory of the loss of Mariner 1 was still strong.

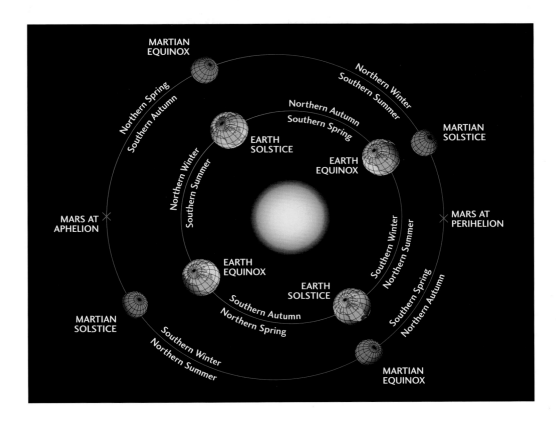

Mars, like Earth, has seasons due to the similar tilt of its axis. Currently, the elliptical orbit of Mars takes it 20 percent closer to the Sun during the southern summer (when that hemisphere is tilted toward the Sun), which results in extremely warm and long southern summers and consequently long and frigid northern winters. Mars travels around the Sun in 687 days, one Mars year, and its orbit is inclined by 1.85° with respect to the ecliptic (the orbital plane of the planets in the solar system).

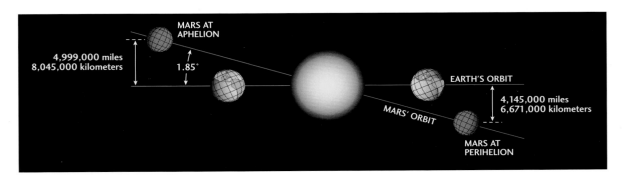

Because of the relative positions of the Earth and Mars, the opportunity to launch a spacecraft to Mars occurs only once every two years or so, near the time when the Earth passes between Mars and the sun and the planets are relatively close together. These encounters are called oppositions, because Mars and the sun are on opposite sides of the Earth. The best time to launch a spacecraft to Mars—the point at which the trip requires the least fuel—occurs about 50 days before each opposition and lasts for two to four weeks. If one of these launch opportunities is missed, space-

flight directors face a two-year wait until the next one.

Mariner 3 was launched on November 5, near the beginning of its launch opportunity. During its ascent, it was protected from the friction of Earth's atmosphere by a protective covering that was supposed to be released when the spacecraft left the atmosphere. Shortly after takeoff, the covering collapsed, crippling the spacecraft. Mariner 3 wobbled out of control. Engineers at the Jet Propulsion Laboratory went on alert, as Murray recalls: "Only three weeks remained to launch Mariner 3's twin, Mariner

4, before Earth and Mars moved out of suitable positions. During those crisis-filled weeks, JPL engineers (with their counterparts at Lockheed Aircraft Corporation, supplier of the Agena rocket) diagnosed why the Mariner 3 shroud had failed, frantically designed a new one, built it, tested it, and installed it on the Atlas/Agena rocket waiting in Florida."

Performing the kind of high-wire act that has become a trademark of the American space program, NASA and JPL engineers succeeded. On November 28, three weeks after Mariner 3's failure, Mariner 4 broke cleanly from the launch pad in Florida and was smoothly maneuvered into a trajectory that would take it to Mars. Just two days later, NASA engineers monitored a secret launch from the Soviet space center. The Soviets did not acknowledge the launch of what they called Zond 2, but NASA's tracking showed that it, too, was on course for Mars. The race to the red planet was on.

Weeks later, disaster struck the Soviets again. Communications with Zond 2 abruptly ceased when the spacecraft approached the location where Mars 1 had disappeared. JPL engineers joked about a "great galactic ghoul" that resided between Earth and Mars, gobbling up spacecraft as they passed. The humor became strained a short time later, when Mariner 4 began to behave strangely while passing through the same region. But the problems were temporary, and Mariner 4 continued to hurtle toward Mars.

Although NASA was working hard to beat the Soviets to Mars, its engineers and scientists had designed a particularly risky mission. JPL's engineers planned to settle the question of the density of the Martian atmosphere—a bit of unfinished business remaining

from the work of Percival Lowell. Mariner 4 was a flyby: it would sweep by Mars in one gentle arc, snapping as many pictures and taking as many scientific measurements as possible before passing out of range. There would be no second chance. But settling the atmosphere question would require a daring maneuver. JPL engineers proposed that the spacecraft be put on a trajectory that would take it behind Mars. That meant the radio signals it was using to communicate with Earth would pass directly through the atmosphere. Careful measurements of the signal would provide a definitive assessment of the thickness of the atmosphere—a critical factor in determining whether the planet could support life.

The difficulty was that when Mariner 4 passed behind the planet, it would be out of sight of the Sun and communication with Earth would be interrupted. Mariner 4's solar panels would become useless, and it would be forced to rely solely on battery power. The spacecraft would be unable to send data back to Earth until after it reappeared. If contact could not be reestablished, Mariner 4's data would be lost. The attempt to measure the atmosphere could put the entire mission at risk. "You had to break radio lock with the Earth, you had to break solar lock with the Sun and go on batteries while you still had all the picture data in the spacecraft," Murray recounted. "So if you lost it, you'd lose the whole mission."

In 1963, however, the need to determine the density of the Martian atmosphere acquired a new urgency. An analysis of reflected light from the Martian surface suggested that the planet's atmosphere might be far thinner than most scientists then believed. The question was urgent because it had a

FOLLOWING PAGES: Mariner 6 image mosaic of heavily-cratered Martian surface suggested moonlike bleakness.

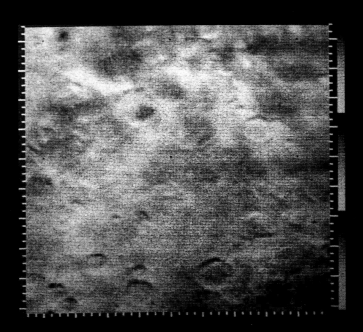

Mariner 4 (above), designed to fly by Mars, had only a brief opportunity to capture images of that planet and transmit them to Earth. After a 7 1/2 - month journey, Mariner 4 passed by Mars on July 14 and 15, 1965. It sent back 21 pictures, including the above image. Taken from a distance of about 8,500 miles, it was the first to show clearly the craters on Mars. In addition to capturing the first detailed images of that planet, Mariner 4's mission demonstrated the feasibility of interplanetary space exploration and made various measurements of space as the craft approached Mars and after passing by. At its closest approach, Mariner 4 came to within 6,200 miles of Mars. Scheduled to have traveled with its sister craft Mariner 3, Mariner 4 made the journey alone. Its traveling companion had been launched a few weeks earlier but had failed just after takeoff.

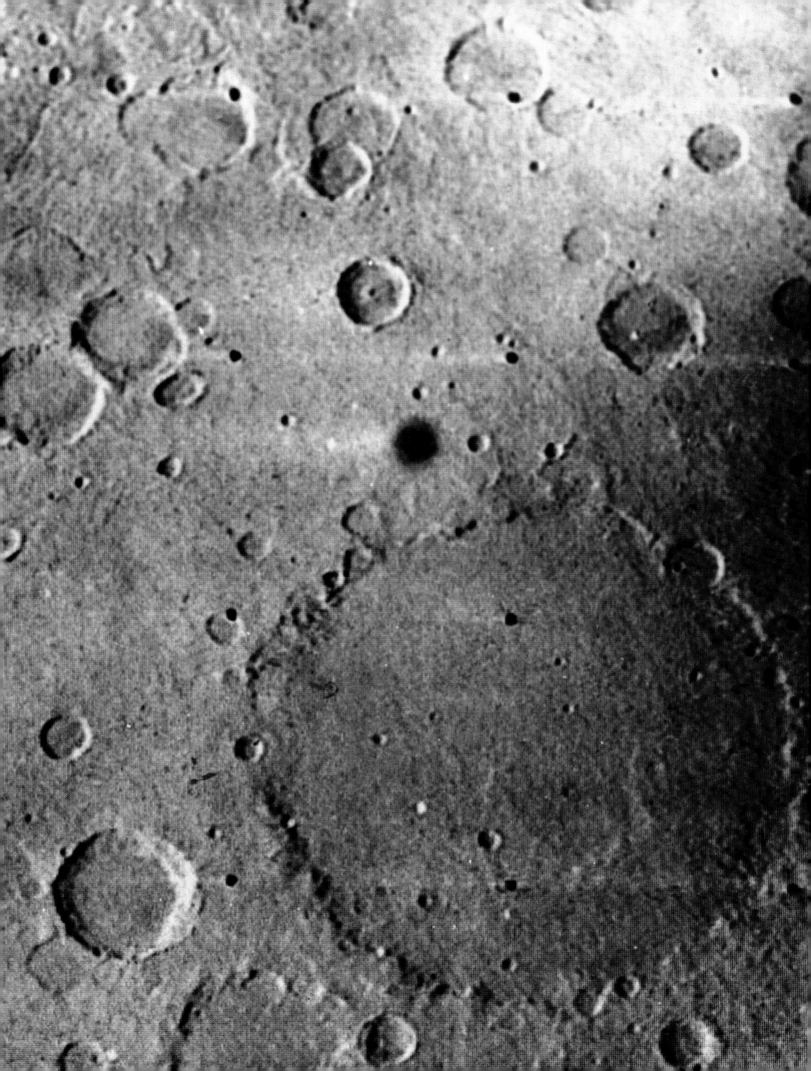

direct bearing on the possibility of life on Mars. If the atmosphere of Mars was extremely thin, any water on the surface would evaporate immediately. Most life depends on the presence of liquid water. If there was no liquid water on the Martian surface, the chance that life existed on Mars would shrink dramatically. The question, then, was whether the new study was correct. Mariner 4, if successful, could put that question to rest. The decision was made to pass behind Mars. "JPL had the guts to do it," Murray said.

On July 15, 1965, more than seven months after it left Earth, Mariner 4 circled behind Mars, becoming the first spacecraft to reach the red planet. America had won the race to Mars. When Mariner 4 emerged on the other side of the planet, engineers picked up its signal again, and it began sending back the first close-up picture of Mars. "Outside JPL's gates reporters from the three major television networks shouted at Frank Colella, JPL's press liaison," Murray recounts. "Percival Lowell, Orson Welles, Ray Bradbury, and a century of Martian speculation and imagery had aroused extraordinary public curiosity about the first close-up views of Earth's sister planet."

Inside JPL's gates, the mood was quite different. Murray and his colleagues sat behind closed doors soberly examining the first image. They couldn't recognize a single feature. Indeed, there was no way to be sure they were looking at Mars. Dust had evidently collected on Mariner 4's camera lens, and the image appeared to be useless. Computers were put to work analyzing the images, trying to subtract the effects of the dust on the spacecraft and haze in the Martian atmosphere. Mariner 4 continued to transmit pictures, many showing the planet in more favorable light. Eventually, to the relief of the JPL team, much clearer pictures of Mars were obtained and released to the reporters clamoring outside JPL's gates. What they saw was as much a shock to the scientific team as it was to the public.

Mars was pocked with craters hundreds of miles across. It was immediately clear that Mars resembled the moon far more than it resembled Earth. This was emphatically not Earth's sister planet. The rugged surface of the planet had apparently escaped anything like the erosion and jostling of crustal plates that reshaped the surface of Earth over the past few billion years. Mariner 4 detected no magnetic field, indicating that, unlike Earth, Mars did not have a churning, molten metal core.

Mariner 4 succeeded in measuring the density of the Martian atmosphere with great precision. And the results came as a crushing blow for those who continued to believe in the possibility of life on Mars. Measurements of the radio waves as they passed through the Martian atmosphere showed that its pressure was about one-half of one percent of that of Earth's atmosphere. That was only about one-tenth what Lowell had calculated—and it was so thin that any liquid water on the surface would instantly evaporate. As it turned out, the Mount Wilson Observatory estimate of the density of the Martian atmosphere was far closer to the truth than previous estimates had been. It was the first step in what Horowitz called the "delowellization" of Mars.

Robert Leighton, the head of Mariner 4's TV camera team, tried to sort out what seemed to be a puzzling inconsistency in the Mariner 4 data. Calculations based on observations from Earth showed that the Martian atmosphere contained about 30 times as much carbon dioxide as Earth's atmosphere. Yet Mariner 4 had shown that the Martian atmosphere was exceedingly thin. The only way to reconcile the two observations was to conclude that the Martian atmosphere was made up almost entirely of carbon dioxide. What would that mean? As Leighton considered this problem, he did some calculations to determine the temperature of the Martian polar ice caps. Their average temperature was about minus 125°C. Suddenly, the significance of that figure hit Leighton and Murray, who was working with him on the study: Minus 125 °C was the temperature at which airborne carbon dioxide and frozen carbon dioxide would remain at equilibrium (that is, frozen carbon dioxide would not vaporize nor would atmospheric carbon dioxide solidify). That could mean only one thing. The polar caps were not

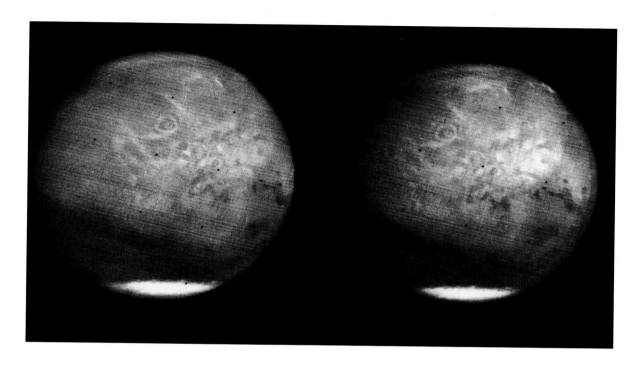

During its approach to Mars in 1969, Mariner 7 shot this pair of images 47 minutes apart, with the south polar cap prominently bright. Note, by comparing the circular Olympus Mons (then called Nix Olympica—not yet known to be a mountain), that in the period between images, the right one shows about 12° rotation, relative to the left.

made of frozen water. They must be made of frozen carbon dioxide—dry ice. One more apparent similarity between the Earth and Mars had collapsed under scientific scrutiny.

The new discovery coupled with that of the crater-scarred surface, delivered a terrible blow to the notion that Mars was a twin of Earth. "There were no Martians, no canals, no water, no plants, no surface characteristics that even faintly resembled Earth's," Murray noted. "In jarring contrast with our expectations, we had just discovered a Mars incredibly inhospitable to life."

Mariner 4 sent back 21 pictures. At its closest approach to Mars, it was less than 6,200 miles from the planet's surface. By the time Mariner 4 had left Mars, it had demolished almost everything scientists thought they had known about the planet. "Before Mariner 4 got there, there was only ground-based telescope data, a lot of which was wrong or misunderstood," Murray recalls. "The error in the knowl-

edge was as large as the knowledge." Before Mariner 4, a variety of things had seemed to mark Mars as a twin of Earth: Mars had nearly the same length of day as Earth, far different from that of any other planet. Its axis had almost the same tilt as the Earth's, far different from that of any other planet. It had similar polar caps that change with the seasons, and the wave of darkening that looked like a seasonal change in vegetation. Before Mariner 4, then, "it was a no-brainer that Mars was a twin of the Earth, that the caps were water frost, which implied liquid water at times on the surface, and that the seasonal markings were primitive plant life," Murray says. "Why would they be seasonal if they weren't tied into the availability of moisture and sunlight?"

As Mariner 4 and subsequent studies showed, scientists had fallen victim to an improbable series of coincidences. The similarities between Earth and Mars were chance occurrences. The length of Earth's day has changed over time, because of the

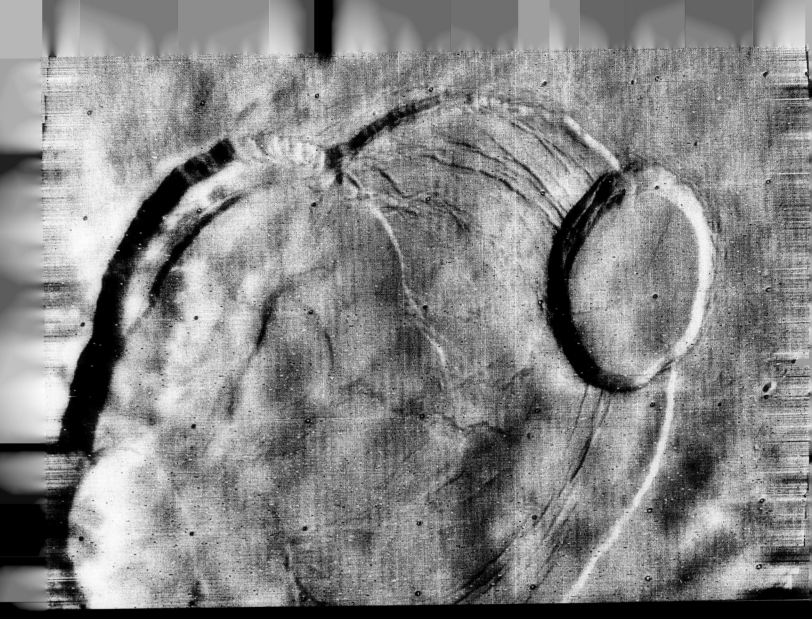

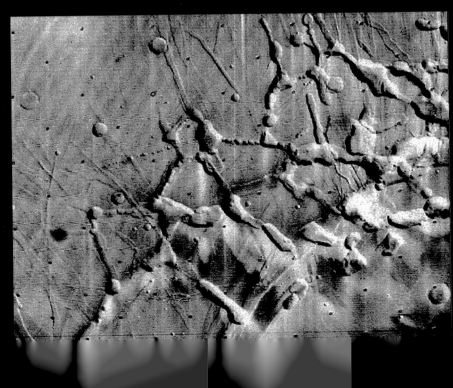

The summit caldera of Olympus Mons, the largest volcano and mountain in the solar system, measures about 100 km across and a few km deep. The scarcity of craters indicates recent geologic activity. The most elevated plain on Mars (right), at the western end of Valles Marineris, shows the Noctis Labyrinthus region of intersecting fault-bounded valleys and erosional troughs.

gravitational interactions with the moon. It is just a coincidence that scientists happen to be looking now, when the day lengths are about the same. The tilt of the axis of Mars also changes over time—and it is just a coincidence that we happen to be looking at a time when it is about the same as the tilt of the Earth's axis. "If we had come out of Africa 100,000 years earlier, it wouldn't have looked like a coincidence," Murray says—the tilt of the Martian axis was far different then. The white of the polar caps is carbon dioxide frost—dry ice, not water ice as on Earth. The possibility of life on Mars seemed finally to have been extinguished. Mars was too cold, its atmosphere was too thin, and its surface was too dry.

The next missions to Mars were ready to go in 1969, four years after Mariner 4. Continuing its practice of sending two spacecraft on each mission, NASA launched Mariner 6 on February 24, and Mariner 7 on March 27. (Mariner 5 was a 1967 mission to Venus.) On July 31, 1969, 11 days after Neil Armstrong became the first person to walk on the moon, Mariner 6 reached Mars, snapping photographs and taking a variety of scientific measurements, precisely as it was designed to do. Mariner 7 arrived a few days later. It was set to explore the polar caps and provide confirmation of Leighton and Murray's conclusion that they were made of dry ice. The spacecraft was behaving normally as it approached the planet when, without warning, its radio signal vanished. Had the slumber of the "great galactic ghoul" been disturbed? Engineers scrambled to reestablish contact with the spacecraft, and they eventually picked up a faint signal—enough to tell them that part of the spacecraft had been destroyed. A rechargeable battery on board the spacecraft had exploded, damaging Mariner 7. Computer commands were sent to Mariner 7, directing it to ignore the damaged instruments and to continue its mission. The commands were accepted, and reliable radio communication was restored. The pulse rates of nervous engineers and scientists dropped back down to something closer to normal, and the mission continued.

The two spacecraft sent back 58 close-up pictures of the Martian surface, more than doubling what had been sent back by Mariner 4. The space probes passed within 2,200 miles of the Martian surface, nearly 4,000 miles closer than the closest approach of Mariner 4. The new spacecraft also carried improved cameras, and the images they sent back to Earth were much clearer. With the new images, researchers had good views of about 10 percent of the planet's surface. (Mariner 4 had shown them only about 1 percent of the Martian surface.) An instrument on Mariner 7 called an infrared spectrometer took a reading of the planet's southern polar cap and sent back an extremely complex stream of data. Leighton, Murray, and a crowd of reporters waited while the data was analyzed. Scientists were astonished at what they had found. Two substances related to life—ammonia and methane—had apparently been detected at the fringe of the southern polar cap. That was precisely where microbes and plants might be found if the cap was made of ice that melted periodically.

"This new argument for life on Mars made headlines all over the world the next day," Horowitz noted. "Had these scientists and newsmen glimpsed the shade of Percival Lowell in the hall?" The excitement did not last long. It turned out that carbon dioxide could itself produce the readings that were taken to be evidence of methane and ammonia. The polar cap was indeed made of frozen carbon dioxide, as Leighton and Murray had concluded. The two spacecraft photographed a few new geological features, but they brought no great surprises. They confirmed the view of Mars that had become established following Mariner 4—that Mars was a brutal, lifeless place. The determination that the polar caps were made of dry ice "became the final nail in the coffin of an Earth-like Mars," observed Murray.

Mariner 6 and 7 were the last of the Martian flyby satellites. Now that NASA had characterized Mars in a general way and had shown how different it was from Earth, it needed a more systematic study of the planet, which would require longer observations. NASA's next objective was to put a pair of spacecraft in orbit around Mars. As the missions were being

The Jet Propulsion Laboratory

Considered the world's premier space science facility, the Jet Propulsion Laboratory leads the Mars exploration program.

In the early days of rocketry, it could sometimes be difficult to explain the difference between scientific research and a Fourth of July celebration. Long before Walt Disney began brightening Southern California with his Magic Kingdom fireworks shows, a handful of would-be rocket scientists at the nearby California Institute of Technology were regularly staging ersatz fireworks shows of their own. More often than not, their experimental rockets would burst into a brilliant shower of sparks and cascading shrapnel. Finally the researchers were chased off Caltech's campus in Pasadena and forced to set up their rudimentary lab in several flimsy shacks at the foot of the San Gabriel Mountains a few miles away.

The big names in rocketry were not part of this group. Robert Goddard, often called the father of modern rocketry, was secretive about his own advances, and he wanted nothing to do with the Pasadena upstarts. (Goddard himself was run out of Massachusetts after his rockets were branded a public nuisance. He moved to an experiment station near Roswell, New Mexico.) Another well-known name is Wernher von Braun. One of the principal developers of Germany's V-2 rocket during World War II, von Braun was also a major figure in American rocketry. In many respects, however, both were overtaken by the small lab established on a shoestring in Pasadena.

That lab would grow to become the National Aeronautics and Space Administration's Jet Propulsion Laboratory, the world leader in the scientific exploration of space.

For more than 60 years, JPL has surprised its critics by accomplishing the impossible. Time and again, the laboratory has taken on engineering and scientific challenges that no one else would dare attempt. Its record isn't perfect, but it has succeeded more often than many outsiders might have predicted. (Its supremely confident engineers would have respectfully disagreed; they knew all along they could do it.) The reward for successful completion of a difficult mission is usually the opportunity to take on a greater challenge—and that often means trying to do more science with less money. When JPL took on the highly successful Pathfinder mission, for example, NASA told the lab it had one chance to do it right; there would be no second Pathfinder. NASA didn't have the time or the money to allow JPL the luxury of wandering into blind alleys. "We were told to take more risks," said Brian Muirhead, one of the leaders of the Pathfinder team, "and not to fail." The seemingly impossible task is the stuff of which the JPL legend is made.

The laboratory was established in 1936 as the Guggenheim Aeronautical Laboratory, California Institute of Technology—or GALCIT. It was nothing more than "a loose band of six amateurs risking their meager Depression-era earnings," according to the historian Clayton R. Koppes of Oberlin College. The leader of this motley group was Theodore von Kármán, a Hungarian-born Caltech professor. One of his students, Frank J. Malina, had been interested in rockets since he was 12 years old, when he read Jules Verne's *From the Earth to the Moon.* Malina found a few others who shared his excitement, and they were soon pooling their meager resources to run a few experiments. In one of them, they accidentally filled a Caltech laboratory with a noxious cloud of methyl alcohol and nitrogen dioxide. In another, a rocket exploded and Malina narrowly missed being hit by a piece of steel that was blasted into the side of a building.

But it was the late 1930s, there was war in Europe, and von Kármán's band soon attracted funding from the U.S. military. GALCIT began doing rocketry experiments for the Army, and by 1944, when it had

been renamed the Jet Propulsion Laboratory, its annual budget was $650,000. As early as 1954, JPL engineers had devised a plan to put a small satellite into orbit. Koppes has concluded that if JPL had been allowed to proceed, it could have launched its satellite by August 1957—two months before the Soviet Union became the first nation to put an object in orbit. The launch of Sputnik was a staggering Cold War propaganda victory for the Soviet Union—one that JPL might have been able to prevent.

By 1958, JPL had turned most of its attention to space exploration, rather than military research. Its Ranger 7 spacecraft was the first U.S. vehicle to hit the moon, and its Surveyor made the first soft landing on the moon. From those early successes, JPL quickly established itself as the leading space research facility in the country, and with a string of failures by the Soviet Union, it became the leading space science laboratory in the world. That is a title it continues to hold.

Frank Malina, third from left, and his colleagues relax and prepare to test an experimental rocket motor in 1936. The hoses supply oxygen and fuel to the rocket, and the square container atop the sandbags holds cooling water.

THE FIRST MISSIONS

planned, few could have guessed what a sensation they would create. Yet again, nearly everything researchers thought they knew about Mars was about to be overturned. The pendulum was about to swing, and the prospects for discovering life on Mars would soon be on the rise.

The next launch opportunity in 1971 would prove to be a watershed. No fewer than five spacecraft were ready to be sent on their way to Mars. The Americans were planning to send two of them—Mariner 8 and Mariner 9. And the Soviet Union had three space probes ready to go. The American spacecraft were orbiters, designed to map and photograph the planet from different orbits. Mariner 8 was designed to fly a polar orbit and map about 70 percent of the Martian surface. Mariner 9's mission was to collect data that might explain the seasonal variations in light and dark areas on the surface—the great "wave of darkening." Slipher and some others believed it might be vegetation. Carl Sagan had speculated that it might be dust blown by seasonal winds; but no one really knew.

The Soviet missions had a more ambitious goal. They would attempt the first landing on Mars. Their space probes were orbiter-landers, intended to put payloads on the Martian soil and send the first pictures back from the surface. Each weighed about 10,000 pounds, more than four times the weights of Mariner 8 and Mariner 9. The Soviets were trying one more time to leapfrog the American effort.

Mariner 8 was the first of this flotilla to be launched. It roared from the launchpad at Cape Kennedy on May 8, only to tumble into the Atlantic a few minutes later when the second stage of its Atlas-Centaur rocket failed to ignite. Two days later, Cosmos 419, the first of the three Soviet spacecraft, was launched into Earth orbit. The last stage of the rocket failed to fire, because an ignition timer that was supposed to be set for 1.5 hours after orbit had mistakenly been set to 1.5 years. The spacecraft fell back into the atmosphere. Following those two failures, the luck of the both the Americans and the Soviets changed. The Soviet space probe

Mars 2 was launched on May 19. Mars 3 was launched on May 28; and the American spacecraft Mariner 9, on May 30. All three were on their way. Six days after Mariner 9 was launched, flight controllers made one mid-course correction. "The maneuver was so accurate that no other correction was necessary for the entire 167-day flight to Mars," flight directors later reported.

When the three spacecraft reached Mars later in November and December of 1971, Mariner 9 turned out to be the star performer. One reason it succeeded was because JPL engineers had made the right decisions during the spacecraft's design. Mariner 9, unlike its Soviet counterparts, was equipped with enough computer power so that its mission could be modified at any time during its flight, even after it reached Mars. That flexibility proved to be crucial.

In September, as the space probes sped toward Mars, a small yellow cloud, visible from Earth, began expanding until it spread across the southern portion of the planet. It continued to expand until it covered the entire planet, obscuring almost everything on the ground and generating punishing winds. The dust storm peaked five weeks after it began, making it the biggest and longest lasting storm that had ever been seen on Mars—and it was occurring at the worst possible moment. For the first time, a tiny contingent of space vehicles was on its way to another planet, and the planet was completely hidden behind roiling clouds of dust. No one could be sure what effect this would have on the success of the missions.

A further complication occurred only days before Mariner 9 was due to arrive at Mars. Engineers were performing a routine flight calibration when the spacecraft's signal disappeared. One of the receiving stations of the Deep Space Network, which tracks interplanetary spacecraft, was able to make intermittent contact. The engineers assumed that the spacecraft had lost its navigational lock on the star Canopus. The spacecraft was designed to automatically begin rolling until it could find the star and lock on again. But this meant that its principal antenna was passing out of sight of receiving stations on Earth with each rotation. Flight controllers averted disaster when they were able to establish communication on a much slower link, using a different antenna. They managed to get the spacecraft locked on Canopus again, and regular communication was restored.

Although Mariner 9 was launched later than the two Soviet spacecraft, it was on a slightly faster trajectory and arrived on November 14, 1971, becoming the first spacecraft to orbit Mars. The Martian global dust storm was raging, and the first pictures sent back from the spacecraft showed almost nothing. About two weeks later, the Russian spacecraft Mars 2 arrived. With a payload bound for the Martian surface, Mars 2 entered Martian orbit. The orbiting spacecraft dispatched its lander, but the lander malfunctioned and crashed into the surface. The lander returned no data, but it did make the Soviets the first to send a spacecraft plowing into the Martian surface, even if the lander was unable to make any measurements or take any photographs when it got there.

Mars 3 followed Mars 2 into successful orbit, and it succeeded in putting a lander on the surface of Mars. The Soviets became the first to accomplish that milestone. Sadly, though, the Soviets' bad luck continued. Seconds after the lander sent a brief signal to Earth, it seemed to disappear. Perhaps it was blown over by the winds driving the dust storm; no one can be sure. Mariner 9 and the orbiting portions of Mars 2 and Mars 3 all remained in orbit around Mars, but the Soviet space probes were locked into a fixed schedule that required them to begin taking pictures. Designers had made no provision for changes in flight. The two spacecraft sapped battery power taking useless images of impenetrable dust clouds (though they did return some useful scientific data). This was where Mariner 9's flexibility became vital to its success. Mission controllers sent the spacecraft a command ordering it to shut off its cameras until the storm cleared, saving power.

About a month after Mariner 9 arrived at Mars, the dust storm began to subside and the surface drifted dreamily into view. Flight directors had changed Mariner 9's orbit to compensate for the loss of Mariner 8. It would now have to do the work of two. Mariner 9 was put into a new orbit that allowed it to accomplish many of Mariner 8's objectives as well as its own goals. That is when the surprises began.

During the dust storm, four strange dark spots had been visible on the surface. As the storm subsided, it became clear that these were enormous craters, as much as 40 to 50 miles wide. What was most unusual about them was that each was perched atop a mountain, an unlikely arrangement for craters thought to have been caused by the impact of mammoth meteorites. No such features had been seen by any of the earlier Mariner flights. At the Jet Propulsion Laboratory, 300 reporters crowded into an auditorium to get the news of the first clear views of Mars from Mariner 9. There the TV image team announced its surprising conclusion: These were not impact craters at all—they were volcanoes. "Our three previous flyby missions had turned up no volcanic features at all," Murray recalled. "Now, suddenly, four such features were the first ones observed by Mariner 9? This seemed implausible to me. But so they were." They were volcanoes far larger than any seen before. The largest, originally called Nix Olympica and later Olympus Mons, is the tallest mountain anywhere in the solar system. It rises nearly 17 miles above the surrounding plains—almost three times as high as Mt. Everest—and at its base it is almost 400 miles across. Its volume is estimated to be 50 to 100 times that of Hawaii's Mauna Loa, the largest volcano on Earth, which rises a mere 5-1/2 miles from its base on the ocean floor. And Olympus Mons was not the only giant Martian volcano that Mariner 9 found. This Brobdingnagian monster stood near three other volcanic giants that were named the Tharsis Montes. Each of them dwarfs anything on Earth.

"And giant volcanoes were just the beginning," Murray said. The next discovery was a vast canyon, the grand canyon of Mars, a huge chasm long enough to reach from San Francisco to New York. Named Valles Marineris, it ranges up to 400 miles wide and six miles deep. Mariner 9 also observed, in the plain known as Chryse Planitia, an intricate, braided

Mariner 9 image, taken on October 12, 1972, revealed the north polar cap, composed of water and carbon dioxide ice. Dark swirls are thought to be frost-free areas. A huge field of dark sand dunes surrounds the cap. Images like this indicate a dynamic climate on Mars with the active exchange of volatiles between the atmosphere and the poles.

network of what looked like channels carved by rivers or floods. What could have caused them? Was there once water on Mars, eons ago?

As Mariner 9 methodically whirled around Earth's supposed twin, it provided a new view of Mars in which neither the Earthlike world of Percival Lowell nor the view of Mars as a barren, lifeless world seemed correct. Mars had apparently had an intensely active history, and most researchers now believe its surface was shaped by catastrophic floods. At some time billions of years ago, the desiccated surface of Mars might have been a churning watery world. Whether that period lasted long enough for life to have emerged is not known, but it is at least possible that some sort of aquatic life or microbial life might have flourished. With Martian evolution then at work, perhaps some of those microbes had found a way to survive.

Mariner 9 findings showed that Sagan was correct about the "wave of darkening." It is caused not by vegetation, but by seasonal dust storms that rearrange the pattern of lighter and darker soils and rocks. The Sinton bands, which were supposed to indicate the presence of lichens and other plants on the Martian surface, also turned out to be a scientific blunder. Researchers showed that the bands were an indication not of plants but of water molecules that included an atom of deuterium in place of one of the atoms of hydrogen. (Deuterium is a naturally occurring form of hydrogen with an extra neutron in the nucleus.) Sinton's bands were caused by deuterium in the Earth's atmosphere—they had no correlation with anything on Mars.

Other findings were puzzling: The northern half of the planet was found to be covered with volcanoes and canyons; the southern half of the planet was heavily pocked with craters; and the braided network of channels in the Chryse Planitia was likely produced by catastrophic flooding, rather than rivers.

Mariner 9 had been designed to operate for 90 days,

but it continued to send information back from Mars for nearly a year. On October 27, 1972, it ceased operation when it ran out of the gas used in the attitude control system. By then, it had sent back 7,329 pictures of Mars and its moons—more data than had been received from all exploration missions in the solar system up to that time. It had become the most important and successful Mars mission in history. "By the end of the Mariner 9 mission, in 1972, our view of Mars had completely changed—once again," Bruce Murray wrote later. "Lowell's Earthlike Mars was forever gone, but so was the moonlike Mars portrayed by our first three flyby missions, Mariners 4, 6, and 7. The Mars revealed by Mariner 9 was not one-dimensional; it was an intriguingly varied planet with a mysterious history. The possibility of early life once more emerged."

The progress in just over a decade had been phenomenal. Carl Sagan explained the explosion of knowledge by comparing it to what had been known before the first spacecraft visited Mars. Before Mariner 4, the fuzzy telescopic images of Mars could be stored in the equivalent of a few thousand digital "bits" of computer information. The 20 photographs taken by Mariner 4 amounted to 5 million bits of information. Mariner 6 and 7 added 100 times as much information, and Mariner 9 increased that another 100-fold. "The Mariner 9 photographic results from Mars correspond roughly to 10,000 times the total previous photographic knowledge of Mars obtained over the history of mankind," he observed.

NASA was already working on its next mission, the most elaborate and expensive mission that would ever be sent to Mars: the Viking project. One of Mariner 9's goals was to select information that would help NASA find the right spot to set a spidery lander down gently on the planet's surface, there to begin the search that Lowell had begun decades before: the search for life on Mars.

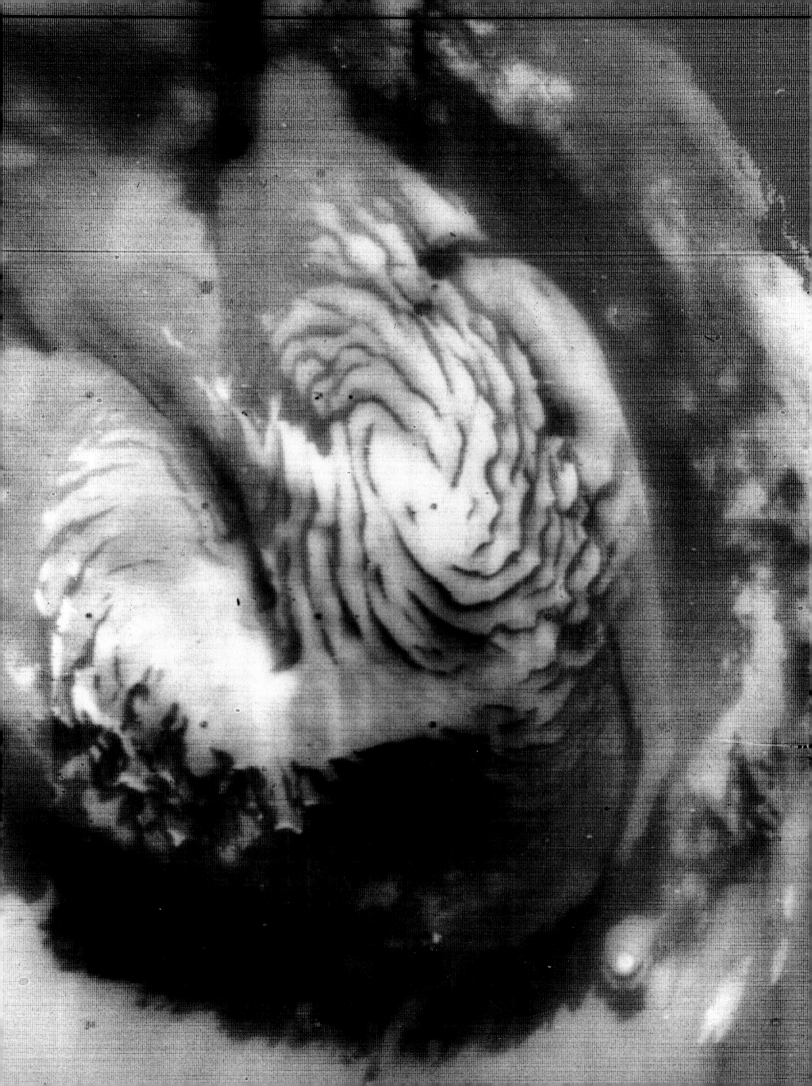

CHAPTER THREE

GETTING TO KNOW MARS: THE VIKING MISSIONS

n 1968, NASA struggled to get the Apollo program back on track in the wake of the devastating launchpad fire on January 27, 1967, that had taken the lives of astronauts Gus Grissom, Edward White, and Roger Chaffee. While the nation awaited the resumption of the race to the moon, NASA scientists quietly continued their work on the Mars exploration program. At the time, planetary scientists were still grappling with the surprising results of Mariner 4—a small pile of photographs of a desolate, crater-pocked world that seemed incapable of supporting life. Mariner 9 would complicate the picture, revealing an active planet, with volcanoes, canyons, and channels apparently carved by catastrophic floods. With those discoveries, the prospects for finding life on Mars—or at least evidence of ancient life—would rise again. But in 1968, Mariner 9 was still three years away. NASA's decision to launch a mission to search for life on Mars was made at a time when few believed there was any life to be found.

The Viking 2 mission to Mars takes off on September 9, 1975, from the Cape Canaveral Air Station, Florida. The Titan III launch vehicle rises on two rocket boosters and the first stage of the rocket.

NASA's new Mars exploration project was called Viking. A daring and extravagant project to land two sophisticated, completely automated biological laboratories on the surface of Mars, Viking's mission was to search for evidence of living things. The Viking flights would be the most ambitious and potentially important scientific missions NASA had ever attempted. The original plan was for the two Viking spacecraft to be launched during the first opportunity after Mariner 9, in 1973. In the summer of 1969, however, plans abruptly changed when Apollo 11 landed on the Moon. On July 20, American astronauts radioed Houston with the words, "Tranquility base here. The Eagle has landed." America had put a man on the moon before the end of the decade, just as President Kennedy had promised. As the world cheered the courage of the Apollo 11 crew and marveled at America's technological accomplishment, Congress prepared to make severe cuts in NASA's budget. President Richard Nixon was eager to support the space program, but decided to restrict the agency's funds until the costly Vietnam War had ended.

The budget cuts meant that the Viking launch would be delayed until the 1975 opposition, although it was unlikely that work on Viking could have been completed in time for a 1973 launch in any event. As it was, its instruments had undergone only minimal testing by the time the spacecraft was launched in 1975, and some of those tests had led to catastrophic results. Even with the two-year delay, some of Viking's scientists worried about instrument failures on Mars.

The launch delay provided the Soviets another opportunity to sweep ahead of the United States. NASA was not ready with a spacecraft for the 1973 opposition, but the Soviets were. Soviet space engineers launched no fewer than four space probes to Mars in July and August of 1973. Their dejected American counterparts could only stand by and watch. By this time, the Soviet space program had sent five successful missions to Venus, including two that landed on the surface and sent back data on surface conditions and soil composition. The Soviets were continuing their policy of pushing ahead with solar system exploration through brute force. One after another, Soviet spacecraft were ejected from Earth's atmosphere on their way to Venus and Mars. Failures were shrugged off. In addition to the five successful missions to Venus, the Soviets had launched 14 that were unsuccessful. Relative to Venus, Mars had proven to be a far more difficult target. None of the four spacecraft dispatched to Mars in 1973—designated Mars 4 through Mars 7—were successful. Mars 4, intended to orbit the planet, was scuttled by a faulty computer chip and never entered Martian orbit. Mars 6, intended to land on Mars, developed trouble during its descent to the surface. As the lander separated from the rest of the spacecraft and began to drop toward the planet's surface, it transmitted data for 150 seconds. When its retrorockets fired to slow its descent prior to landing, contact was lost. The Mars 7 lander separated early and missed the planet altogether. Mars 5 orbited, returned data on the atmosphere, and sent back images of a small part of the Martian surface. But it operated for only a few days, and then its transmissions stopped. Despite a bold attempt, the Soviet Union had missed its chance to gain ground on the American planetary exploration program. By the time of the next launch opportunity, in 1975, NASA was ready with its twin Viking spacecraft.

Each Viking spacecraft carried 14 instruments intended to determine the composition of Martian soil, to study the atmosphere and climate, and to infer something about the planet's interior. But above all, the Viking spacecraft were designed to search for life. Some scientists criticized the decision, arguing that the probability of life on Mars was too small to justify such an enormous commitment of money, time, and scientific and engineering talent. The Viking project was going to cost one billion dollars. The debate over the huge cost continued for a time, but it was soon overtaken by a disarming practical question. How, exactly, would Viking search for life? Biologists could offer no simple, indisputable test to determine whether life was present. The Viking landers would be equipped with cameras and would

Artist's rendering depicts the final descent to the Martian surface of a Viking lander more than 20 years ago. The retro-rockets used to slow the lander raise surface dust as it lands on three legs with footpads. Viking landers searched for signs of life in the Martian soil and discovered none. Surface temperatures proved too cold and pressure too low for liquid water to exist. Without water, required for life as we know it, the prospects for finding life on Mars today seem remote.

The Viking missions to Mars comprised two identical spacecraft, each having a lander (left) and an orbiter. Each lander sat atop the orbiter within a white pod (above, top). The orbiter has four solar panels as well as instruments that include the orbiting cameras (on the scan platform at right above the solar panels). On the left of the orbiter the antenna relays lander data to Earth. The lander, with antenna on top, scoops up soil for sophisticated life-detection experiments (left and above, close-up). The lander's white boom has meteorology sensors; its white cylinders contain cameras. Because of the lander's shape and the positioning of the cameras, they could view the directly adjacent surface only in the vicinity of the sample arm.

easily be able to identify anything that walked by or sniffed around them. Still, it was easy to imagine circumstances in which large, readily visible creatures might be missed. "I keep having this recurring fantasy that we'll wake up some morning and see on the photographs footprints all around Viking that were made during the night, but we'll never get to see the creature that made them because it is nocturnal," remarked astronomer Carl Sagan, one of the most enthusiastic proponents of a search for life on Mars.

Sagan wanted a flashlight on Viking, but mission planners turned him down. He also suggested that Viking should put out bait. The idea was to put various nutrients on the outside of the spacecraft. In his book *The Search for Life on Mars*, Henry S. F. Cooper, Jr., recounts the slightly goofy debate that ensued, as Sagan insisted that he was serious about a search for large, ambulatory Martians that might meander over to the spacecraft and lick nutrients off of it. (These creatures were referred to as "macrobes," to contrast them with the microbes that were being sought in Viking's biology experiments.) "Since it's so cold on Mars, I don't expect anything to be nocturnal," said Joshua Lederberg, a Nobel Prize-winning biologist and friend of Sagan's who was then at Stanford University and later went on to become president of The Rockefeller University. "It would be coldest at night, and most macrobes would be asleep then, trying to conserve their warmth."

Sagan disagreed. "Some animals on Earth, such as Arctic fish, have elaborate antifreezes," he said. "...If you're a predator, what better device would there be than to go around at night with a heat sensor, eating Lederberg's sleeping organisms?" Others noted that creatures able to withstand extreme cold are uncommon even on Earth, but Sagan was undeterred. "Someone has to propose ideas at the boundaries of the plausible, in order to so annoy the experimentalists or observationalists that they'll be motivated to disprove the idea," he said.

The Viking missions were actually made up of four spacecraft: two that would orbit Mars and two landers. Set for launch in August and September

1975, they were to be put into a gently curving trajectory that would bring them alongside Mars in the summer of 1976, after a journey of almost 500 million miles. The plan was to set the Viking 1 lander down on the surface on July 4, 1976—the nation's 200th birthday. Viking 2 was scheduled to land about two months later.

The orbiters were designed to photograph the planet, to measure the amount of water vapor in the air, and to study atmospheric temperatures. Each orbiter carried two high-resolution television cameras and two instruments to make atmospheric and surface observations. The landers carried instruments for use during their descent to measure the makeup of the Martian atmosphere and to monitor its temperature, pressure, and density. Once the landers had reached the surface, these instruments and others would study Martian biology—if it could be found—and the planet's chemical structure, weather, and earthquake activity. The landers were also equipped with cameras to photograph the Martian surface.

Although the search for life was Viking's primary goal, the project had other scientific objectives, some of which were designed to answer questions raised by the 7,000 pictures returned by Mariner 9. The study of the atmosphere was important not only because of what it could tell researchers about Mars, but because of what the results might tell researchers studying Earth's atmosphere. It is difficult for climatologists on Earth to make generalizations about atmospheric science when there is only one planet's atmosphere to look at. The information from Mars would provide researchers with a second planetary atmosphere to compare with Earth's. A particular focus of the Viking spacecraft was to discover whether there was any water on Mars. "It is known that there is water in the Mars atmosphere, but the total pressure of the atmosphere (about one percent of Earth's) will not sustain any large bodies of liquid water," said Viking's chief scientist, Gerald A. Soffen, and several colleagues in a review of the mission. "Nevertheless," they continued, "the presence of braided channels suggests to many geologists that they are the result of previous periods of flowing water. This...suggests a very dynamic planet."

The exploration of Mars, Soffen explained, was part of a larger effort to understand the formation and history of the solar system. At the time of the Viking missions, he listed some of the key scientific questions: What are the characteristics of the Martian surface? What are the constituents of the thin, vaporous Martian atmosphere, and how did it get that way? How does climate change on Mars? What is the internal structure of Mars, and what is its history?

More specifically, volcanologists wanted to find out what was happening in the interior of Mars that caused the string of giant volcanoes along the Tharsis ridge. Likewise, geologists wanted data that might help them explain what caused the huge Martian canyon, Valles Marineris, that was discovered by Mariner 9. And the atmospheric studies, researchers hoped, would help explain why there had been no indication of nitrogen on Mars. Had it been lost, or was it locked up in the soil?

All of these investigations were secondary, however. "There is the final question of life on Mars," Soffen and his colleagues wrote. "This may be one of the most important scientific questions of our time. It is also one of the most difficult to answer." If the Viking landers failed to find life, that would not prove Mars is barren; it would simply leave the question open. Despite nearly a century of speculation about life on Mars, there was still no evidence of life when the Viking missions were being planned. "What we seek is evidence," the researchers wrote. "The remarkable thing is that we live at a time in which we can make this first test for life."

Why is the search for life on other worlds so compelling? The discovery of life on Mars could reveal much about the origin of life, one of the central questions of biology. Many of the details of evolution remain unclear, and the discovery of another example of evolution on another planet could vastly increase understanding of that fundamental process. Such a discovery could reveal, for example, which characteristics of life are common to life everywhere, and which are particular characteristics of life on Earth. Is it possible that other organisms could evolve in ways entirely unlike they did on Earth, or is Earth an example of what is likely to happen anywhere? Without the discovery of life on another planet, researchers cannot know. Beyond the scientific interest, however, lies a deeply rooted human curiosity about life on other worlds. From the time of the ancients who named the stars and the constellations, people have speculated about the possibility of life somewhere out there. Never before have they had any hope of finding out.

The biological experiments aboard Viking were the subject of enormous controversy. Indeed, they continue to be the subject of argument among a tiny group of researchers who believe—despite overwhelming opposition from other scientists—that Viking might have found something out there. Nothing like Viking's search for life had ever before been attempted, and researchers needed to think very carefully about the nature of life before they could begin to design experiments. The researchers differed among themselves about what they should be looking for.

While Sagan worried about the issue of nocturnal Martian bipeds wandering by Viking when the lights were out, the real job facing Viking was far more difficult: determining the presence of organic molecules—which make up life on Earth—or microbes in the soil. But what would such microbial life be like? Terrestrial life is based on carbon. The organic chemicals that are the principal components in living things are all based on carbon, which has an unusual ability to form a huge variety of compounds. Some scientists speculated that life elsewhere might be based on silicon as an alternative to carbon. How would Viking search for silicon-based life? Of course, if Martians turned out to be big enough to be spotted by Viking's cameras, it wouldn't matter what they were made of. A silicon-based creature who wandered by would be as easy to spot as one made of carbon or anything else. But the search for microbes would be far trickier. An automated laboratory

Carl Sagan

A distinguished astronomer, the late Carl Sagan was also known as the foremost proponent of the search for life on other planets.

As a boy in Depression-era Brooklyn, Carl Sagan took his first library card to the local branch library and asked for books on stars. The librarian quickly suggested books on Hollywood stars, but Sagan, after a few minutes of embarrassment, explained that he meant the stars in the sky. When he was given an astronomy book, he was astonished to learn that the stars were just like the Sun, only very far away. And he learned what the planets were. "Then it seemed absolutely certain to me that if the stars were like the Sun, there must be planets around them. And they must have life on them...

I thought of it before I was eight."

The idea never left him. Sagan went on to become a distinguished astronomer, a Pulitzer Prize-winning author and host of public television's *Cosmos* series in 1980. Above all, though, Sagan was the driving force behind the search for life on other worlds. He combined a rigorous scientific skepticism with a wondrous imagination and childlike fascination with the possibility of life on other planets. As a member of the Mariner and Viking satellite teams, Sagan, who died at the age of 62 on December 20, 1996, probably had as much influence on NASA's Mars exploration program as anyone.

Sagan was as disdainful of alien-abduction and UFO enthusiasts as he was of scientists who seemed almost to wish that life would not be found. "Some people are nervous about the possibility of life on Mars, even simple life," so that, faced with two alternatives, they prefer to "choose the least interesting" one, Sagan once wrote. When others said there was no evidence to support the existence of life on Mars, Sagan would turn that around: There is no evidence that excludes the possibility of life, he would argue. At the same time, he had little patience with the credulous accounts of people who claimed to have been visited by aliens or to have communicated with them. His success and influence in the scientific community rested on the dual pillars of his rigorous scientific skepticism and his boundless scientific imagination and willingness to follow the data wherever it led.

On occasion, he followed the data to scientific realms far removed from the search for life. In the 1980s, Sagan and a handful of colleagues sounded an alarm about what they named "nuclear winter," the idea that a nuclear war could inject such a great quantity of fine smoke particles into the atmosphere that the world would undergo a rapid, catastrophic cooling. Such a climate shift could, in the extreme case, make the Earth unsuitable for human life. The theory arose from Sagan's consideration of data from the Mariner 9 spacecraft. Monitoring Martian temperatures during a ravaging global dust storm, Mariner 9 found that surface temperatures had dropped sharply. "Over the subsequent twelve years, this line of inquiry led from dust storms on Mars to volcanic aerosols on Earth to the possible extinction of the dinosaurs by impact dust to nuclear winter. You never know where science will take you," Sagan wrote in his 1994 book *Pale Blue Dot.*

Sagan studied astronomy at the University of Chicago and taught at Harvard before moving permanently to Cornell University in 1968. He published more than 600 scientific papers and was the author, co-author, or editor of more than 20 books, many of which were best-sellers. He wrote the greeting from Earth that was carried aboard the Pioneer 10 spacecraft, and he was involved in the search for signals from space that could be evidence of extraterrestrial life. His *Cosmos* television series was seen by more than 500 million people in 60 countries, making Sagan arguably the greatest science popularizer ever. He was the recipient of numerous academic honors, including the Public Welfare Medal, the highest award of the National Academy of Sciences. In making the award, the academy said "his research transformed planetary science," and "his gifts to mankind were infinite."

Sagan often sparred with Bruce Murray, the distinguished geologist at the California Institute of Technology who also worked on the Mariner and Viking missions. Murray was far more skeptical about the possibility of life on other planets. "I find it difficult to maintain the optimism of a [Ray] Bradbury or a Sagan," Murray said after the Mariner 9 mission. "Since Sagan and I respect each other greatly as scientists and find much stimulation in each other's thoughts, why should we find it so difficult to read the record similarly?" he wondered, and he supplied an answer: "My first love was, and is, the Earth," he said. "Carl has emphasized synthesis and conjecture about how things are, might be, or could be beyond the Earth," Murray said.

Although the two researchers disagreed on many things, they were both passionately committed to space exploration, and they ultimately found themselves working together. With Louis Friedman, who worked on space research at the Jet Propulsion Laboratory in the 1970s, Sagan and Murray founded the Planetary Society to encourage space exploration and support related educational and scientific activities.

Murray once said that if Sagan was lucky, "his great passion, the search for extraterrestrial life, especially intelligent life, [would] blossom during his lifetime." Sagan died before unequivocal evidence of life on other worlds was found, but his writings and research in the field of exobiology—the study of life on other planets—are likely to influence America's space exploration efforts for years to come.

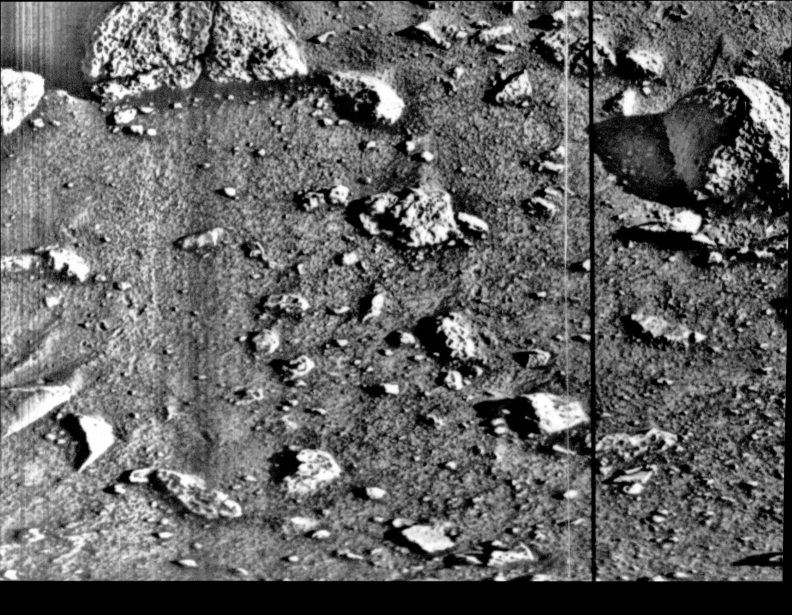

First picture taken from the surface of Mars by Viking Lander 1 on July 20, 1976, shortly after landing, was transmitted slowly, line by line to Earth. The image shows the footpad on a rock- and soil-covered surface. The footpad has fine-grained material within it, deposited during landing. The large rock in the center of the image measures about 4 inches (10 cm) across. Viking 1 landed in Chryse Planitia, a low area near the equator into which flowed many catastrophic

floods. The northern third of the planet, less heavily cratered and lower than the southern two-
thirds, indicates that geological processes
resurfaced and lowered the northern plains early
in Martian geologic history. Sitting astride the
highland-lowland boundary, the huge Tharsis
Bulge contains four giant volcanoes. Young volcanics associated with Tharsis cover about one
quarter of the planet's surface, and tectonic
features extending out from Tharsis influence the
entire western hemisphere of Mars.

Panorama of Martian surface taken from the Viking 2 lander in Utopia Planitia (above) shows the lander's arm at lower center and the white meteorology boom extending upward. The surface of Mars here appears as a fairly flat and homogeneous angular rock field. Images such as this Viking orbiter mosaic of channels (right) in the heavily cratered terrain of Mars argue for the action of liquid water in the early history of this planet.

FOLLOWING PAGES: Mosaic of over 30 individual Viking orbiter images shows dividing, subdividing, and reuniting channels in the region of Mangala Vallis. The channels drain a heavily cratered surface and required liquid water to have eroded them.

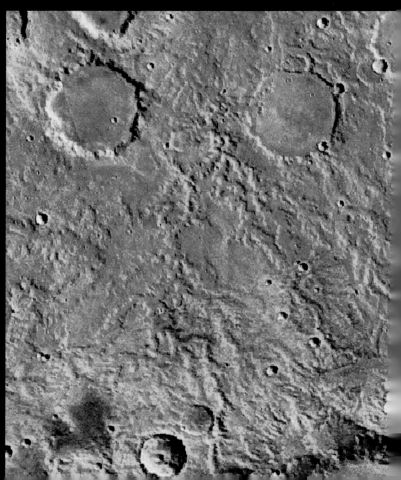

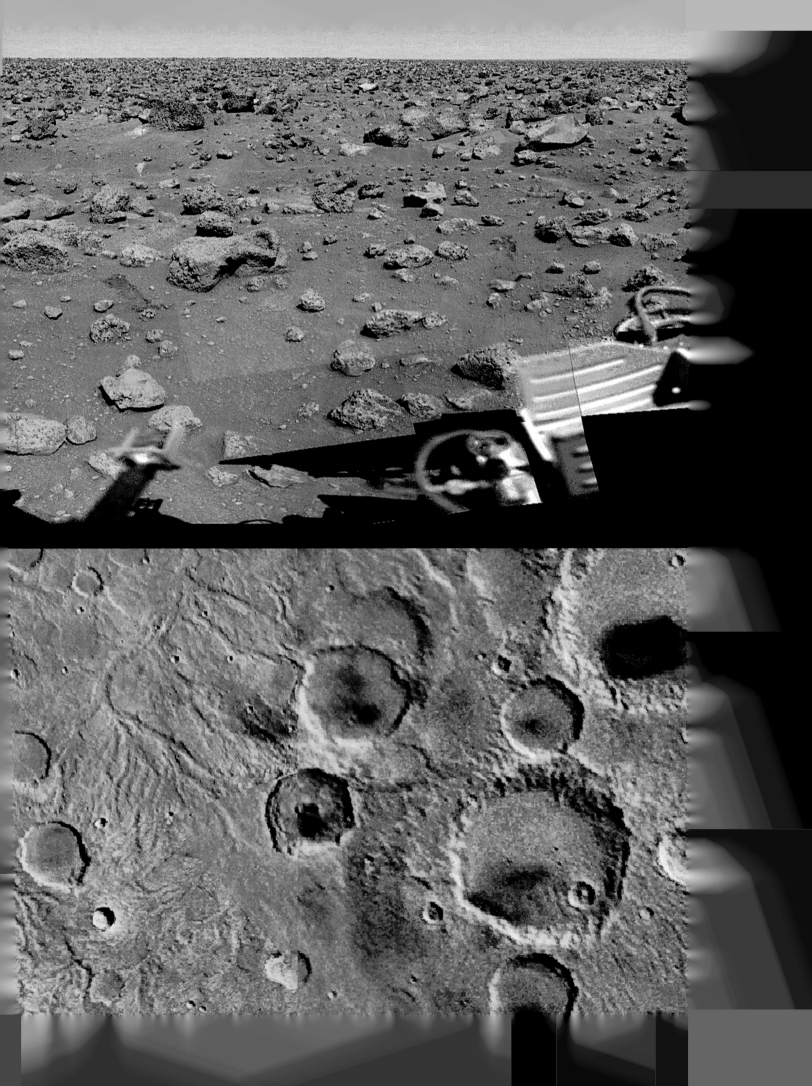

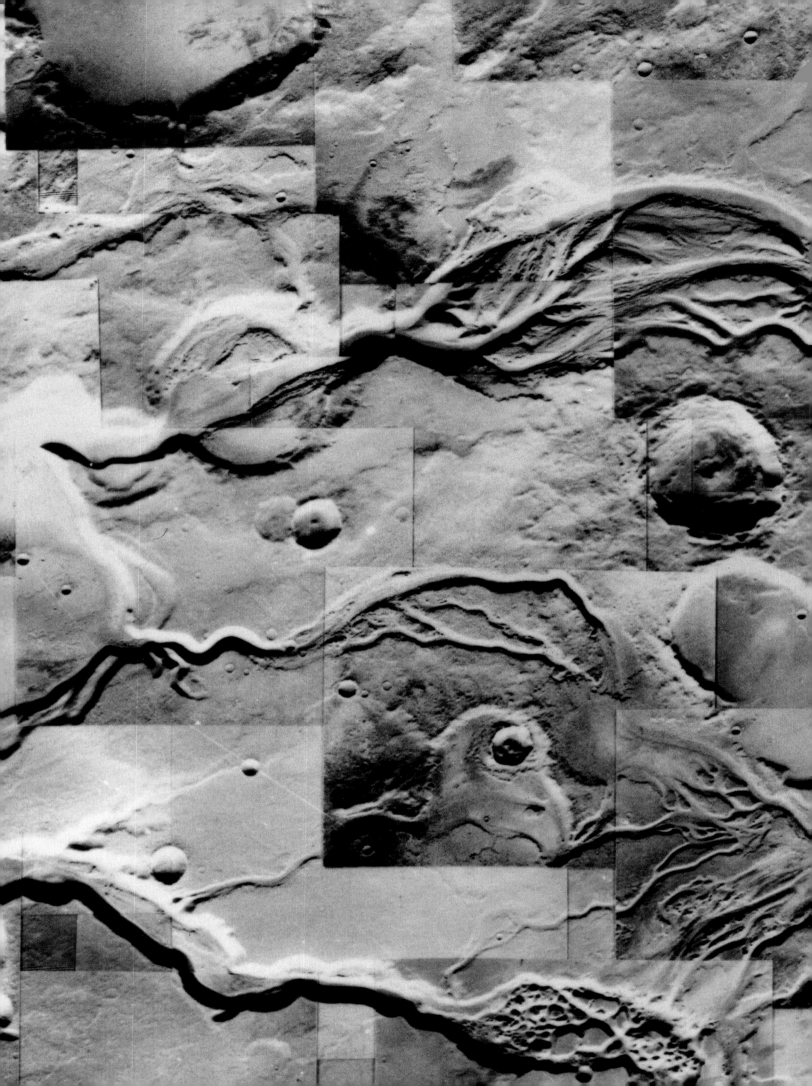

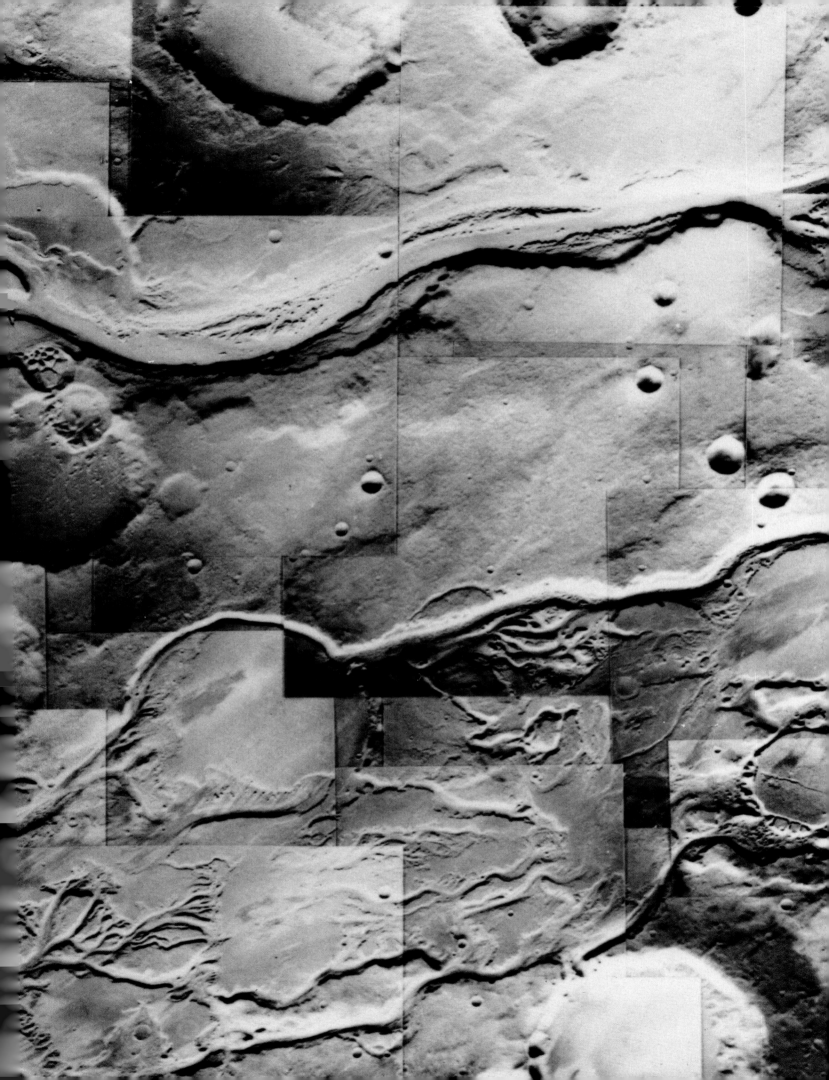

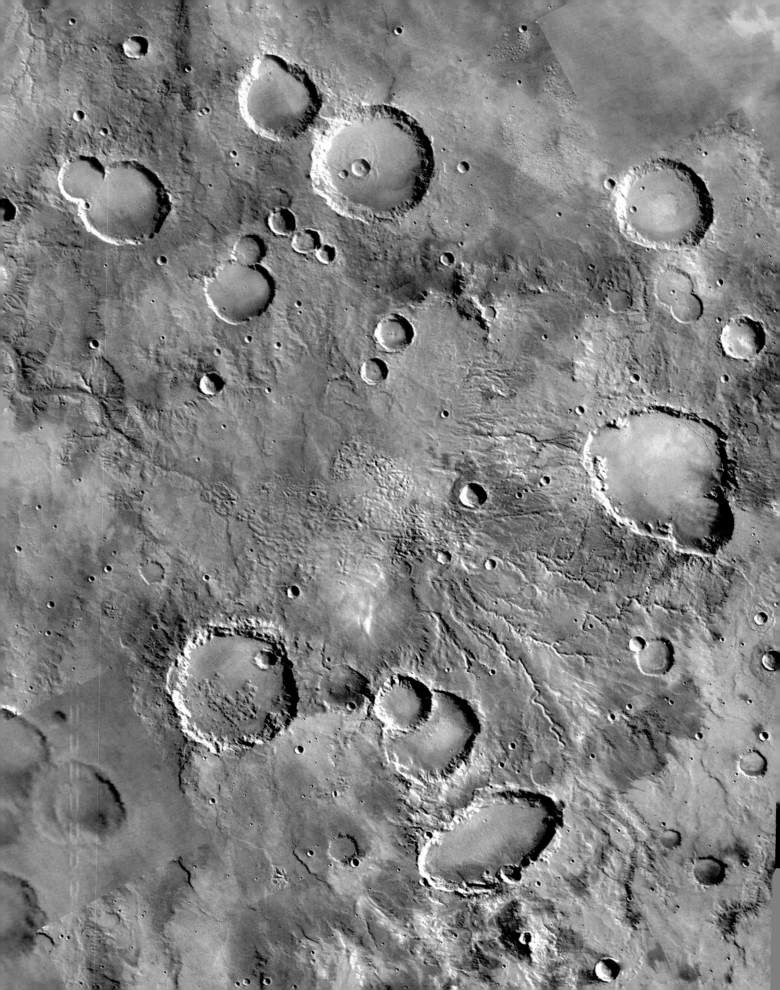

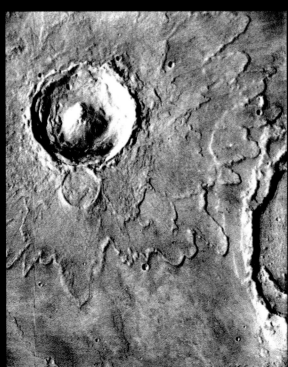

This Viking image mosaic oblique view (above) of Ravi Vallis illustrates a classic chaotic region from which catastrophic floods originated. Interpretations suggest that water welled up out of the ground, producing the region of jumbled blocks and the escarpment. By analogy with catastrophic outflow channels on Earth, huge volumes of water must have carved the channels. Crater Yuty (left), with a central peak, probably had ice-rich crustal materials that flashed into water during the impact that formed the crater. A Viking mosaic of the Margaritifer Sinus region of Mars (opposite) shows a cratered terrain with fine channels ending at a low area in the center, once a lake of water, suggesting that water may have existed in equilibrium with the environment early on in Mars's history. If so, conditions may have supported the formation of life.

The edge of the uplifted and rough Argyre impact basin, its relatively smooth interior, and the Martian atmosphere on the edge of the horizon show up in this oblique Viking orbiter view. Predominantly carbon dioxide, the atmosphere of Mars has a surface pressure only a hundredth that of Earth. This low pressure and the cold surface temperatures negate the possibility of liquid water (now stable under these conditions as a solid or gas only) on the Martian surface.

designed to detect organic chemicals might completely miss clues left by more exotic organisms.

The key factor in the search for life was the presence of water. Earth's living organisms have proven to be extremely resilient, able to survive huge extremes of temperature, thriving in the perpetually black ocean depths and the wind-scoured snows of Antarctica. But in every case known on this planet, living things need water. It is essential for the biochemical reactions organisms need to make energy and survive. The Earth is a marine planet—the salty taste of a human teardrop is a reminder that life originated in the seas billions of years ago. But the surface of Mars was freeze-dried billions of years ago. Could some sort of waterless, solid-state life have developed there, unlike anything we know? Viking's scientists certainly couldn't rule that out. But they found it difficult to imagine how the chemistry on which life is based could take place without water or some other solvent. The job of the Viking designers was to try to imagine what life on Mars might be like, and to design experiments to find it. This would be by far the most ambitious space science research ever attempted. If it succeeded—if it found life on Mars—it would have a cataclysmic effect on human thought, changing humanity's view of itself forever. Less than 20 years after the first satellites had been lofted into wobbly orbits around the Earth, researchers were busy devising a way to search for life on another world. The progress had been remarkable.

Carl Sagan was by far the most imaginative and optimistic of the Viking scientists. He repeatedly provoked and sometimes angered his colleagues with his irrepressible optimism about life on Mars. But he also thought more deeply than many of his colleagues about how the search for life should be conducted. Certainly, if Viking's television cameras saw footprints one morning that hadn't been there the day before, that would be a pretty good indicator of the existence of life. If the comparison of a series of pictures of the same area suggested that something was moving, that would be another fairly obvious clue. But Sagan wanted more than that; he wanted a careful definition of what constituted life, a set of yardsticks that could be used to determine whether something was alive. For example, one such definition of life could be: Something that consumes energy. Or something that moves. Or something that can reproduce. But Sagan felt those definitions were all inadequate for evaluating findings from Mars. "All those old definitions don't really fit," he said. "My automobile eats and breathes and metabolizes and moves. Crystals grow and even reproduce." Sagan finally settled on a criterion he referred to as thermodynamic disequilibrium—that is, a departure from what was expected. Anything that seemed to require energy—a plume of heat, say, or an object seeming to remain out of balance—might be an indication of life, by Sagan's new definition. A cow, Sagan noted, is an improbable thing: its big, bulky body is supported on four spindly legs. It is top-heavy, and it ought to fall over—which is exactly what happens when it dies. The reason a cow maintains its top-heavy position, improbably balanced on thin legs, is that it is alive.

Another scientist who spent a great deal of time thinking about how life might be detected was Norman Horowitz of the California Institute of Technology. Horowitz, a member of the biology team for the Viking missions, noted that life has two key features. First of all, living things are far more complicated than nonliving things. And second, living things appear to have a purpose, as they work to ensure their own survival. The growth and behavior of living things is governed in part by their genes, a

Viking mosaic of a portion of Valles Marineris reveals Ophir (top) and Candor Chasmata, which are about 61 miles (100 km) across and about 6 miles (10 km) deep. Top of mosaic shows the cratered plateau surface and the steep face of

walls) giving way to hummocky landslides in the
upper right and left. Smooth material, forming
plateaus within the troughs, are interior layered
sedimentary deposits, suggesting their having
formed in lakes and therefore arguing for standing
bodies of water relatively recently on Mars.

phenomenon not seen in nonliving things. Genes are intimately involved with reproduction and evolution, both of which are critical indicators of life. As Sagan pointed out, automobiles and other machines can mimic some properties of life, but automobiles don't carry a genetic code. Nor do they evolve; they rust. As Horowitz put it, "Now we can answer the question 'What is life?' The genetic attributes of living things—that is, the capacities for self-replication and mutation—underlie the evolution of all the structures and functions that distinguish living objects from inanimate ones....Life is synonymous with the possession of genetic properties. Any system with the capacity to mutate freely and to reproduce its mutations must almost inevitably evolve in directions that will ensure its preservation. Given sufficient time, the system will acquire the complexity, variety, and purposefulness that we recognize as 'alive'."

Horowitz had provided an instructive way to think about life, but what did that mean for Viking? At the time of Viking's launch, researchers had just learned how to isolate genes on Earth, and it was only with painstaking work using the new tools of genetic engineering that they were able to identify them. That wasn't going to work on Mars. Nor was there anything Viking could do to demonstrate the existence of evolution on Mars. Horowitz's conclusion was that any life that might be found on Mars would almost certainly be made of organic molecules— substances containing carbon atoms. "Carbon is so superior for the building of complex molecules that the possibility of forming genetic systems with other elements has never seriously been considered," Horowitz said. It is unlikely, he argued, that silicon molecules, for example, would allow the mutations and replication necessary for evolution to take place.

"Some may find it disappointing and perhaps even a little depressing to conclude that the surest way to find life on another world is to search for complex chemical systems based on carbon," Horowitz wrote. But there is little hope, he continued, "for exotic creatures formed from, say, vanadium, molybdenum, or praseodymium." That is not to say that life on Mars could not take exceedingly exotic forms. The thing that makes carbon the most likely component of life—its ability to combine in myriad ways with other elements—is precisely what could allow strange and wonderful creatures to develop somewhere else, even if they were based on carbon. Furthermore, carbon is one of the most abundant elements in the universe, Horowitz noted, and "abundant elements are more likely to be involved than rare ones."

Horowitz attached great value to Viking's cameras. As unlikely as it was that they would capture images of Martians, they were the only instruments aboard Viking that could provide absolute, unequivocal confirmation of the existence of life. The great shortcoming with the cameras was that they would not be able to see anything less than several millimeters across. In the absence of the discovery of life, the cameras were likely to send back the most accessible and engaging data. Geologists and physicists who might profess a preference for cold, digitized data tend to congregate around monitors, watching, transfixed, images arriving from another world.

Each of the Viking landers carried six instruments to search for life. The array of instruments included, most importantly, two television cameras. Each lander also carried an exquisitely sophisticated instrument called a gas chromatograph-mass spectrometer, or GCMS. This piece of machinery, designed by Klaus Biemann, an organic chemist at the Massachusetts Institute of Technology, was intended not to find life itself, but rather the constituents of life— organic compounds. Most of the organic compounds on Earth were produced by living things, but organic molecules can be produced in the absence of life. Meteorites are known to contain certain organic compounds. Other organic substances are thought to be closely linked to living organisms. The GCMS would be able to distinguish among various organic compounds. That meant it should be able to distinguish evidence of life from evidence of other kinds of organic chemicals.

In addition to the two cameras and the GCMS, each lander carried three biology experiments. These would test soil samples directly for indications of microbial growth. Each was designed to operate in a different way. Collectively, they were designed to provide the best possible chance of picking up any trace of life, any scrap of information suggesting that something alive had been dropped into one of the tiny test chambers. The Viking biology experiments would become the center of spirited debate, prompting some sharp disagreements among the scientists trying to interpret what they found. Although only a scattered few researchers would claim that Viking found any evidence of life, some of the questions raised by the biology experiments remain unanswered.

By the time of the Viking mission, Mars was known to be highly unsuitable for life. The average temperature of the Martian surface is about minus 64°F. The average temperature on Earth, in comparison, is about 60°F. Nighttime temperatures on Mars are far lower. At noon on Mars, in sunshine unobscured by dust storms or clouds, temperatures can reach as high as 65°F. The extremes would not necessarily preclude life. The more serious problem was the lack of water and the thin atmosphere.

On Earth, all living cells, with the exception of some that go into a dormant state, depend upon water. Human blood is a fantastically complex mix of cells and molecules, but from the standpoint of its water content, it is almost indistinguishable from distilled water, according to Horowitz. (In technical terms, the two have nearly identical "water activity.") That is why hospitals use a simple solution of salt water to increase fluid in patients' bloodstreams.

Horowitz considered whether extreme environments on Earth might provide clues to the possibility of life on Mars. Desert plants, for example, might help explain how plants could survive on an arid, wind-whipped Martian plain. It turns out, however, that desert plants require nearly as much water as plants that thrive in wetter environments. Desert plants have not evolved schemes to survive with less water; instead, they have evolved complicated strategies to conserve the little bit of water available in the desert. Some have extremely short life cycles that allow them to grow and reproduce during the very short seasons when water is available. Such plants leave behind dormant bulbs or seeds to sprout during the next wet season. Other desert plants survive on water vapor or tiny bits of condensation during cool evenings. But even in the most parched desert environments on Earth, water vapor is present in amounts "that are enormous by Martian standards," Horowitz noted.

If desert plants cannot serve as models of life on Mars, what about desert animals? As with plants, most desert animals need almost as much water as other animals, but they have unusual ways of getting it. Horowitz considered the case of the kangaroo rat, found in Arizona and California. It manufactures water from carbohydrates. Laboratory experiments have shown it can live without water indefinitely if given barley, but its ability to manufacture water falls off when humidity drops below 10 percent—which is still far above the humidity on Mars. Desert insects can extract water vapor from the atmosphere, but, again, they can do so only at humidities far higher than those found on Mars.

The Earth environment most closely resembling Mars is that of the Antarctic deserts. These dry valleys have an average temperature of about minus 5°F. Even in the Antarctic summer, the average temperature is only about 32°F., right around freezing. Life in these valleys is mostly limited to microbes—molds, yeasts,

FOLLOWING PAGES: White areas in the north polar region of Mars, interpreted as ice-covered, contrast with finely layered whitish and reddish layers, perhaps clean and dirty ice. Because these surfaces appear very young (almost no craters), the layers may record short-term climatic changes or fluctuations. The 1998 Mars Polar Lander, planned for launch in January 1999, will land on the south polar layered deposits, equipped with instruments to investigate the layers.

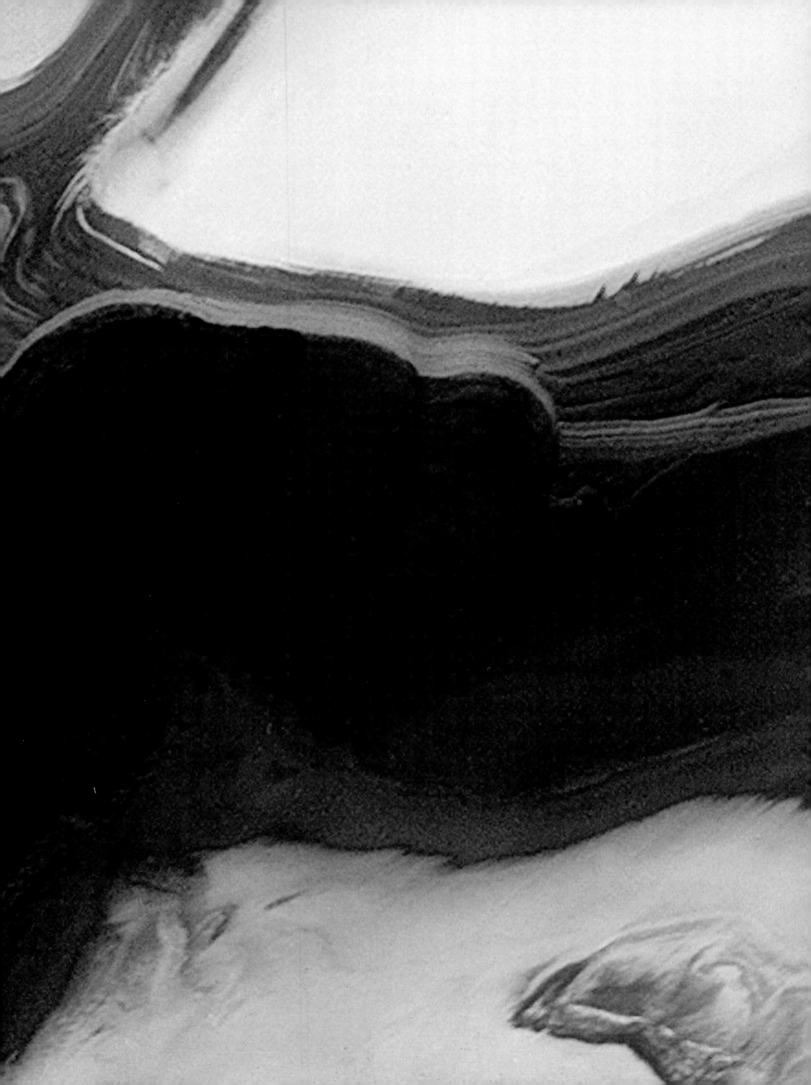

Early morning panorama of a portion of the Viking 1 landing site in Chryse Planitia reveals drifts of fine-grained materials to the left of the white meteorology boom. The large rock on the left, Big Joe, measures more than three feet (about a meter) across. This site—more heterogeneous than Viking 2's landing site—has drift-covered and rocky areas, at right. Floods, impact, and underlying outcrop may have been factors in the site's origins, still under debate.

bacteria, algae—and even those are scarce. Living things have a difficult time hanging on in the dry valleys, the most hostile environments on Earth. Even these areas, however, are more suitable for life than the surface of Mars, which is far colder and drier.

Horowitz's survey of extreme environments on Earth led him to a pessimistic conclusion about the possibility of life on Mars. "Although the adaptations of desert organisms to life with limited water are ingenious and often surprising, they become almost irrelevant when measured against the extreme dryness of Mars," he observed. "Among all known terrestrial species, only water-vapor utilizing lichens can even be considered as possible models for Martian life....If life exists on Mars, it must operate on different principles from terrestrial life."

In addition to lacking water, Mars lacks anything like the Earth's ozone layer to protect living organisms from the biologically harmful effects of the sun's ultraviolet rays. On Earth, the ultraviolet rays are blocked by the ozone layer. The scientific and environmental concern over the thinning of the ozone layer and the appearance of ozone "holes" at various times during the year arises because such damage to

the ozone layer allows more ultraviolet radiation to reach the Earth. On Earth, increases in ultraviolet radiation can produce adverse effects on human health. In 1969, Mariner 6 and Mariner 7 confirmed that the surface of Mars is bathed in a lethal dose of radiation every time the sun rises. The only way microbes could survive on Mars, it seemed, was if they were underground, where they would be shielded from this deadly assault. In recent years, however, researchers on Earth have discovered life in extremely hostile environments, such as fiery hot springs and volcanic undersea vents. It no longer seems so clear that the rugged conditions on Mars would preclude the existence of life.

Sagan, while acknowledging these realities, continued to imagine scenarios in which life of various kinds might find protected niches where it could surmount the apparent hazards of the Martian surface. And for a time, he had at least one enthusiastic ally. Wolf Vishniac, a microbiologist at the University of Rochester and a member of the Viking biology team, believed that the dry valleys of Antarctica must harbor more life than most scientists believed. Vishniac had worked since the late 1950s on the development

of a simple test to look for microorganisms. The device, referred to by his friends as the Wolf Trap, was initially chosen to go to Mars aboard Viking. In 1971, however, budget cuts forced Viking managers to cut back from four biology experiments to three. They decided that the Wolf Trap would be cut from the mission. "It was a crushing disappointment for Vishniac, who had invested 12 years in its development," Sagan said. "Many others in his place might have stalked off the Viking Biology Team."

Instead, Vishniac decided to pursue his research in the next best place—the dry valleys of Antarctica. It was Vishniac's belief that no one had done a proper search for life in Antarctica. During the austral, or southern, summer of 1971-72, he went to the dry valleys, using new tests that he believed would reveal the presence of life where none had been found before. He tried to grow microbes from the Antarctic soil, using a nutrient soup that differed from what had been used previously. The initial results during that first summer led Vishniac to believe he was on the right track. He returned to Antarctica the following year. On December 10, 1973, he went to a dry valley in the Asgard mountain range to collect sampling devices he had left there a month earlier. "It was the last time anyone saw him alive," Sagan recounted sadly in *Cosmos*. "Eighteen hours later, his body was discovered at the base of a cliff of ice."

Vishniac had apparently wandered into an unex-plored area, where he slipped on the ice and fell 160 yards to his death. "Perhaps something had caught his eye—a likely habitat for microbes, say, or a patch of green where none should be," Sagan wrote. "We will never know." Some of Vishniac's microbiology samples were later brought back from Antarctica and examined, and a variety of microbes were found, including a new species of yeast. Some were found a millimeter or two inside rocks, where small quantities of water had been trapped. If the findings made after Vishniac's death had been incorporated into the design of the Viking biology experiments, or if Vishniac's Wolf Trap had made the trip to Mars, would they have changed the outcome? Perhaps the answer to that awaits some future Mars mission. For Horowitz, however, this is wishful thinking. "The question of life in the dry valleys may still be an open one for some," he wrote. "The relevance of the answer for Mars, however, is no longer at issue. The Mars question was settled by the Viking mission." Not everyone agrees with that view today, but many scientists felt that way at the conclusion of the Viking mission.

While the debate over the biology experiments continued, NASA's Viking planning team was busy readying the biology experiments and figuring out how to transport them safely to the surface of Mars. The scientists and engineers were concerned with building the spacecraft, launching it, getting it on the proper trajectory to Mars, putting the landers down gently on the surface, and giving the experimental devices the commands to begin. Much of the planning work was similar to that for the Mariner spacecraft and other NASA planetary voyages. But with Viking, two novel and difficult problems had to be considered. The first was the selection of landing sites for the two spacecraft. Mission planners wanted sites that were free of rough terrain, rocks, or crevices that could topple or swallow the landers as they touched down. The biologists, however, had some ideas about the best places to look for life, and they wanted their concerns taken into consideration, too. The second problem for Viking was what to do about the risk of contaminating the Martian environment with microbes from Earth. Without special precautions, some of the ubiquitous microbes on Earth could easily hitch a ride on Viking, perhaps surviving to reach the Martian surface, where they might take hold and flourish. The likelihood that terrestrial microbes could survive in the hostile Martian environment seemed remote, but the mere possibility was chilling: Would the people of Earth, on their first visit to Mars, bring along microbes that would overrun and destroy the pristine Martian environment? If indeed there is life on Mars, could contamination with microorganisms from Earth destroy it?

The choice of landing sites was made against the backdrop of the rather unsettling history of Soviet attempts to land on Mars. The Soviets had tried many times to land a spacecraft on the Martian surface. All had failed. The Soviets had, however, landed five spacecraft on Venus. Was there something about Mars that made landings difficult? The Soviet spacecraft Mars 3 had landed during a storm, as the surface was being scoured by extremely high winds. The Soviets had the longstanding disadvantage that their flights were incapable of being modified once they had begun. The mission planners for Mars 3 did not have the flexibility to delay landing or change the landing site to avoid the raging storm. Viking, like other NASA spacecraft, would have options. Each

spacecraft would be put into orbit around Mars first, so it could observe its prospective landing site closely before dispatching its lander to the surface. The choice of landing sites was based upon decisions made after examining Mariner 9 photos. When the Viking spacecraft arrived in orbit around Mars, their cameras would be able to take much more detailed photographs of the chosen landing sites before the landing attempts were made. If unpromising features were revealed by the orbiters' cameras, the choice of landing sites could be changed. Furthermore, if either landing site was found to be in the midst of a storm, the landers could be sent elsewhere or kept in orbit until the stormy weather subsided.

The landing sites could not be too close to the poles, because long, cold nights could strain the landers' equipment. In addition, sites too close to the poles would have only a short period each day in which they could communicate with mission controllers on Earth. The landing sites had to be relatively smooth and free of debris. The soil had to be hard enough to support the spacecraft, but not so hard that the Viking mechanical arm would be unable to scoop up soil to test for the presence of life. Mariner 9 images of the Martian surface did not show features smaller than about 100 yards across. The Viking orbiters, while providing better images, would still be unable to see any boulders that might be scattered on the landing sites. No matter how careful mission planners were, they would be forced to contend with an element of luck.

One bit of help came from radar images of prospective landing sites. Rough sites or very soft sites scattered the radar beams rather than reflecting them back to the source, so these undesirable sites looked darker than smoother, firmer sites. "There were many constraints—perhaps, we feared, too many," Sagan, who participated in the decision, wrote afterwards. "Our landing sites had to be not too high, too windy, too hard, too soft, too rough, or too close to the pole. It was remarkable that there were any places at all on Mars that simultaneously satisfied all our safety criteria. But it was also clear that our search for safe harbors had led us to landing sites that were, by and large, dull." The tentative landing site for Viking 1, at latitude 21°N, was located in an area called Chryse Planitia, where channels presumably carved by catastrophic water flows empty into the northern lowlands of Mars. The site for Viking 2 was at latitude 44°N in Cydonia. The site was chosen because there was reason to think it might have a bit of liquid water at certain times of the year. The Chryse site could not be scanned with radar until a few weeks before landing, because of the relative positions of Earth and Mars, and the Viking 2 site at Cydonia was too far north to be seen with radar at all. "I found myself making very conservative recommendations on the fate of a billion-dollar mission," Sagan said.

As it turned out, both landing sites were later changed. After the radar scan and a close examination by Viking 1 from orbit, the Chryse site seemed too risky. The photographs from the orbiter could only give clues, not answers. Any boulder larger than about eight-and-a-half inches could damage the lander, but the Viking orbiter could not distinguish rocks of that size, or even come close—it couldn't see anything less than 100 yards across. Even with such poor resolution, however, the images from orbit showed that the original landing site was covered with carved channels. Radar images from Earth were studied to try to find an alternative site. "That was quite a circus," David Pieri, who was part of the mapping team, said later. "We'd get the new data at 10 o'clock each evening, and we'd stay up all night preparing a new site map to present to the mission management."

It took scientists a month of reconnaissance to find another site in Chryse that was acceptable. The delay upset NASA's plan to have Viking 1 touch down on July 4, 1976, when the nation was celebrating its bicentennial. Another site in Chryse was found. Likewise, the Cydonia site was soon abandoned for

Viking 2. Photographs from orbit revealed craters, boulders, and a network of enormous cracks. Viking 2 landed, instead, at a place named Utopia.

Sagan was often an irritant to scientists trying to identify landing sites. He seemed to come up with objections to almost everything proposed. He worried about the possibility of winds disturbing the Viking landers as they touched down and questioned whether dust on the planet's surface might be deep enough to allow the landers to sink out of sight. Thomas Mutch, the leader of the lander imaging team, had a hard time imagining all the problematic scenarios that Sagan kept envisioning, according to Henry S.F. Cooper, Jr. "I look at the pictures of Mars from Mariner 9, and I say the ground looks good," Mutch told Cooper. "Sagan says that the resolution is poor—only one-hundred-meter objects show up. I say, 'Yes, I know that, but I can extrapolate from what I know of the Earth and the moon.' He says 'Yes, but can you really be sure?' And, of course, I can't."

When he wasn't busy raising questions about the selection of landing sites, Sagan was worrying about the possibility of contaminating Mars with microbes from Earth. During the Apollo program, Sagan argued for extensive quarantining to prevent the contamination of Earth with any microbes that lunar astronauts might bring back with them. Efforts to seal the Apollo astronauts in protective suits and quarantine chambers for a period of time had not worked well. Although the moon proved to be lifeless, and therefore there was no epidemic of lunar microbes, Sagan often reminded Viking scientists of what might have happened. He argued that any mission to bring rocks back from Mars raised the possibility of contamination of Earth with Martian microbes. While such a mission is under consideration for early in the 21st century, Viking was not designed to return with surface samples.

FOLLOWING PAGES: Enhanced evening color view of a portion of the Viking 1 landing site discloses the red drift material and soil covering Big Joe, at left. The oblong drift-covered rock called Whale (note dark tail at left) rests near the rear center of the image. Portions of the white meteorology boom and the lander show up at lower right and lower left.

At one point during the Mariner 9 mission, scientists considered shrinking the spacecraft's orbit so it could get a closer look at the Martian surface. Sagan and Joshua Lederberg vigorously objected, on the grounds that the spacecraft might crash if it got too close, littering the surface with terrestrial microbes. That didn't happen, but the mere possibility shaped later discussions about Viking. As a consequence, the two Viking landers were heated for 80 hours before launch at temperatures as high as 233°F. The process took place while the spacecraft were sealed inside containers called bioshields, which would remain in place until the spacecraft were on their way to Mars. Indeed, the United States and other nations had drafted a treaty in which they pledged to do everything possible to avoid "harmful contamination" of extraterrestrial bodies.

Horowitz thought the difficulty in finding life in Antarctica's dry valleys made it clear that the risk of contaminating Mars was exceedingly small. If microbes could barely survive in Antarctica, they would never make it on Mars, which was far colder and drier. Nevertheless, NASA decided that to fulfill the obligations of the treaty, the Viking spacecraft would have to be sterilized. That added to the cost of preparing the spacecraft, and it risked damaging the instruments. Imagine, for example, putting a television set inside an oven for more than three days at temperatures above that of boiling water. That was what the far more sensitive electronic devices on Viking had to withstand.

Viking's search for microbial life depended on four instruments—the three biology experiments and the gas chromatograph-mass spectrometer, or GCMS. The job of the GCMS was not to search for life itself, but for the carbon-containing organic molecules on which life is based. Organic molecules do not definitely signify the presence of life. The organic matter on Earth is virtually all produced by living things, but that is not true in space. Meteorites contain organic molecules formed through chemical processes, and the guess was that the surface of Mars would be littered with organic material from meteorite impacts even if no life existed there.

The GCMS, like the biology experiments, was to be given a sample of soil dug up by Viking's retractable arm, which had a reach of ten feet. Soil scraped from the surface would be ground, passed through a sieve and delivered to a tiny oven, where it would be heated to more than 900°F. That would spring the organic material free from the soil and break large organic molecules into their constituents. The vaporized molecules would pass into a long column designed so that different molecules passed through it at different speeds. That was the gas chromatograph portion of the GCMS. As each of the molecules exited the column, it would go to the mass spectrometer, where it would be broken down further by an electron beam and passed through electrical and magnetic fields. Such manipulations would allow researchers on Earth to identify the molecules. The device was extremely sensitive—it was capable of detecting nearly all organic compounds in concentrations as low as a few parts per billion.

The three biology experiments aboard the Viking landers each took a different approach toward the detection of life. One of the experiments, called the pyrolytic release experiment, was designed by Norman Horowitz. The idea was to put a sample of Martian soil into a small chamber containing Martian atmosphere and radioactive carbon monoxide and carbon dioxide. The sample would be left in the chamber for 120 hours. The Martian atmosphere is primarily made up of carbon dioxide, and any microbes that live on Mars have likely learned to use it. If such microbes were present in the chamber in Horowitz's experiment, they should pick up some of the radioactive carbon Horowitz had added. The idea then was to flush out the atmosphere, heat the sample to release the organic material and see whether it contained any radioactive carbon. If it did, that would provide strong evidence of the presence of life.

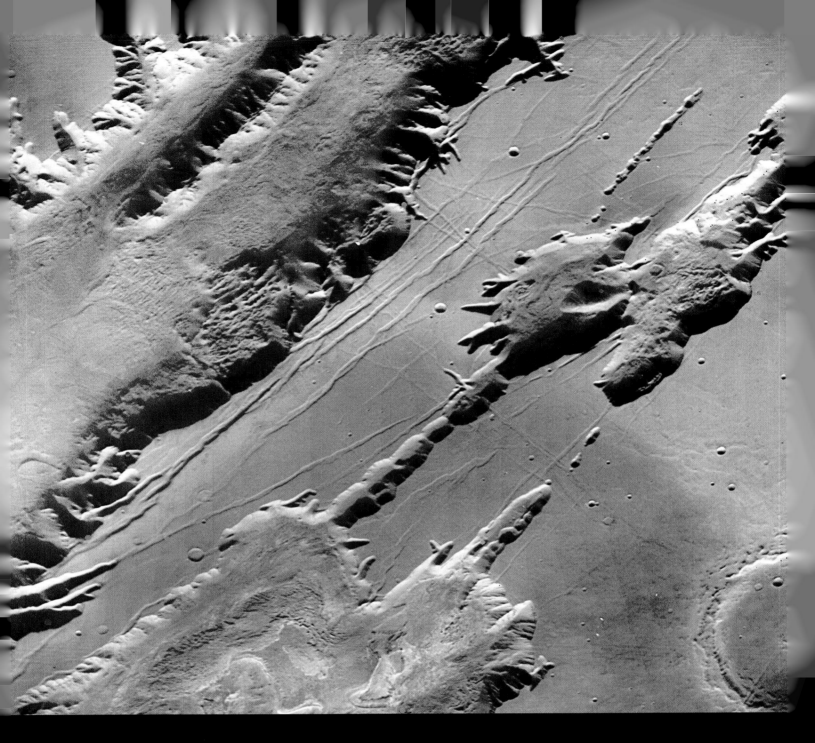

Viking orbiter image about 62 miles (100 km) across
depicts different topographies of the canyons of Coprates
Chasma of Valles Marineris. At the top left lie plains with
steep, linear canyons. The troughs toward the lower right
display effects of erosion, including rounded edges. The
relief is huge—many times greater than the Grand Canyon.

Fear & Flight – The Moons of Mars

The year 1877 was an especially good year for Mars watching—the opposition, or close pass, that year brought Mars to within 35 million miles of Earth, as close as the two planets ever get. In the United States, an American astronomer, Asaph Hall, took advantage of Mars's close proximity to challenge the widely held belief that Mars had no moons.

In early August, Hall began scanning the sky around Mars, using one of the best telescopes in the world at the time—the 26-inch refractor at the U.S. Naval Observatory in Washington, D.C. Hall soon spotted several faint objects. Each turned out to be a distant star. He devised a way to look at the sky very close to Mars without letting the glare from Mars reduce the view, but by August 10th he had found nothing.

His wife, Angeline, urged him to continue, and he tried again the next night. Shortly after 2 a.m., he recorded the observation of a faint star near Mars, but rising fog blocked his view. The next several nights were cloudy, and it was not until August 16 that he was able to manage another glimpse of the faint object he had found. Watching its motion, he realized that it was not a star. He had found what he was looking for. A moon. On August 17, waiting for it to reappear, he found a second moon.

He named them Phobos (Fear) and Deimos (Flight) after the two attendants of Ares, the Greek god of war—like the Roman Mars—in Homer's *Iliad*.

Phobos and Deimos are unlike Mars itself. Oddly shaped lumps of dark rock entirely unlike the red rock of Mars, they are so small and reflect so little light that astronomers were unable to learn much about them until the era of spacecraft exploration. Phobos, the closer moon, orbits about 3,700 miles above the Martian surface—so close that Mars, seen from Phobos, would be a dazzling red globe filling nearly half the sky. About 17- by-13 miles in diameter, Phobos is small

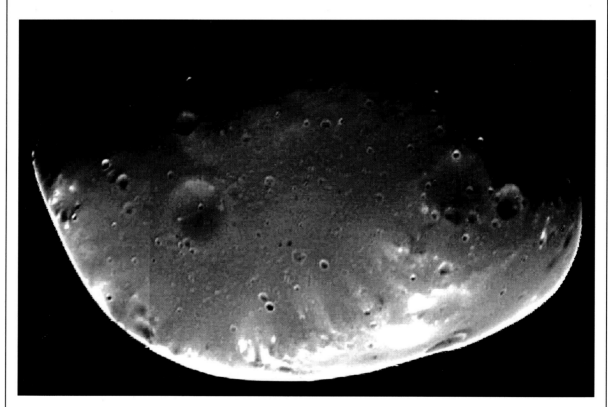

Deimos, the smaller of Mars's two moons, orbits the planet at a mean distance of about 14,500 miles. Its surface appears smoother than that of the moon Phobos (opposite).

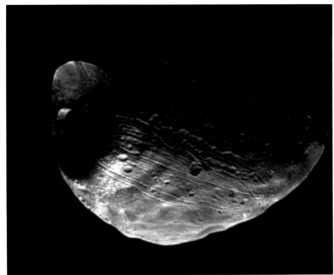

Phobos is gouged by the huge Stickney Crater, named for Asaph Hall's wife, Angeline Stickney Hall.

enough for a reasonably fit hiker to circumnavigate in a few days. Its gravitational pull is so weak that an astronaut standing on the surface could hurl objects fast enough for them to escape into outer space. Deimos, the outer moon, is even smaller, measuring about 10-by-6 miles in diameter.

Before exploratory spacecraft reached Mars, astronomers had determined that Phobos, the inner moon, was accelerating slightly, meaning that it could plunge into the Martian surface in less than 100 million years. Deimos, on the other hand, is gradually pulling away from Mars and might, one day in the distant future, escape entirely. In 1959, a Soviet astrophysicist, Iosif S. Shklovskii, studied the data on the motion of Phobos and concluded that it must be exceedingly light—only one one-thousandth the density of water. Wondering how it could possibly be so light, he concluded that it must be hollow. Phobos might be, he proposed, a space station launched by some long-since vanished Martian civilization.

It was an enchanting notion, one sure to have roused the ghost of Percival Lowell. But the first close-up photos of Phobos demolished the idea. Mariner 7 took a picture from a distance of 82,000 miles, revealing that Phobos was a lumpy, scarred piece of rock, not the shiny, metallic hull of a space station. After Mariner 9 returned three dozen photos of Phobos and

Deimos, scientists said Phobos looked more like "a diseased potato." A better description yet might be a diseased potato with a bite out of it—one crater on Phobos is about one-third Phobos's size. Deimos, similarly, was found to be irregularly shaped and pocked with craters.

The curious appearance of the moons raised an immediate question: Where did they come from? One theory is that the Martian moons are asteroids that became caught in Mars's gravitational field. The motion of the moons, however, suggests they originated in nearly circular orbits not too different from their present orbits. That observation prompted another explanation for the moons' origin—they could have been built up over time by the gradual accretion of material that remained after the formation of Mars.

Yet another puzzle is why the surfaces of the two moons are so different. Phobos is covered with craters resembling those found on the Moon. The craters on Deimos are less prominent and filled with debris. And Deimos lacks the grooves that are such a striking feature of Phobos. Yet the makeup of the two moons is quite similar.

An opportunity to resolve some of these questions came in 1988 when, after eight years of planning, the Soviet Union launched the Phobos 1 and Phobos 2 space probes to orbit and explore their namesake. The two spacecraft were launched in July 1988, but both failed to return enough data to resolve the puzzles about the origin of the moons or their relationship to each other. "We have more data and better crystallized ideas," Joseph A. Burns of Cornell University later noted. "But they lead in different directions."

GETTING TO KNOW MARS

Horowitz's experiment had the virtue of looking for life under Martian conditions, in a Martian atmosphere at Martian temperatures. The other two experiments searched for life using nutrients in solutions of water. Horowitz was critical of the others because they essentially recreated Earthlike conditions to look for life. In order to keep the nutrient solutions liquid, the experiments had to be heated to higher temperatures than what Martian microbes would be accustomed to. And the experiments would expose any Martian microbes to a healthy dose of liquid water—something that doesn't exist on Mars. The concern on the part of some was that these experiments would either fry any Martian organisms they encountered, or drown them.

One of the more Earthlike biology instruments was called the gas-exchange experiment. It was designed by Vance Oyama of NASA's Ames Research Center at Moffett Field, California. In that experiment, Martian soil would be exposed to a nutrient broth at a temperature of about 50°F. The gases in the chamber would be monitored, using a gas chromatograph (like the one in the GCMS). The idea was to see whether the composition of the gases changed in any way that might indicate the presence of life. The third biology experiment, designed by Gilbert Levin of Biospherics, Inc. in Rockville, Maryland, was called the labeled-release experiment. As with the gas-exchange experiment, it depended on exposing soil to an Earthlike nutrient solution. It differed from Oyama's experiment in several respects, however. It used much simpler nutrients, the kind that are known to exist in meteorites and interstellar clouds. And instead of using a gas chromatograph to get the results, it used radioactively labeled carbon, as did Horowitz's experiment. That made it very sensitive. And unlike Oyama's experiment, which mixed soil with nutrients, Levin's labeled-release device would simply drop a tiny bit of nutrient solution on the soil, rather than immerse the soil in solution. Horowitz gave Levin's experiment a left-handed compliment by pronouncing it "an ideal life-sensing device for an aqueous planet"—which Mars, of course, was not.

Before Viking was launched, the biology experiments and the GCMS would have to be tested. In order to interpret the results from Mars, the biology team wanted to expose the instruments to a variety of circumstances on Earth, to learn exactly how the experiments were likely to behave on Mars. That would make the researchers more confident of their interpretations of the results that would soon be coming back from Mars. The problem was that the biology instruments were already far over budget. Each was supposed to cost about $18 million. The actual cost was triple that. There was no money left for testing. Horowitz, Oyama, and Levin had built highly imaginative, sophisticated instruments— and now they would have no time to learn to use them.

Some earlier testing had been conducted, but it had been mostly disastrous. In 1973, a test of the carousel that rotated the different experimental chambers under the Viking soil scoop was a complete failure. The carousel got stuck, instrument readings went off of the charts, and certain metal parts of the biology package were bent out of shape. In another test only seven months before launch, the instruments were tested in a vacuum chamber simulating atmospheric pressures on Mars. The experiments seemed to be working well until soil was put into them. Only Oyama's gas-exchange experiment worked. After the test, technicians found a plugged gas line and broken plungers. The package was fixed and tested again. And, for a second time, only Oyama's worked. When the scientists examined the apparatus this time, they found that one of the chambers had exploded when put in a vacuum, spewing dirt all over the entire biology package.

The GCMS had troubles, too. In a test in which there were no organic chemicals in the instrument, it returned readings showing that it was detecting organic chemicals. Scientists found that pieces of a grinder used to break up the soil were getting into the soil. The grinder was made of carbon steel, and the carbon in the steel was finding its way into the test chamber, where it gave rise to the false readings.

The problem was corrected in time, but with little margin for error.

Despite the minimal testing, Viking's planners hoped that the combination of the three biology experiments and the GCMS would give them a clear reading about the presence or absence of life on Mars. If the experiments found no life, Sagan and other determined optimists would surely come up with scenarios to explain how life on Mars might still exist, even if Viking hadn't found it. And if Viking seemed to indicate the presence of life, Horowitz, Bruce Murray, and a few other skeptics could be counted on to come up with alternative explanations of the findings. The planning and construction of Viking had taken eight years, and during that time an estimated 10,000 people had played some role in the project—planning, designing, building, and testing the spacecraft and preparing to guide them to Mars. The 13 science teams—responsible for designing Viking's experiments, operating them, and interpreting the results—included 1,000 scientists. With all that effort, NASA planners felt they had foreseen most eventualities, and they were hopeful that the results of the biology experiments, in particular, would be illuminating, whether or not they found evidence of life. What Viking's planners and scientists could not possibly have anticipated, however, was that the surface of Mars would be far different from what anyone had guessed. The exploration of Mars had been full of surprises, and Mars had yet another surprise in store.

The launch of Viking 1 was delayed for nine days by a series of nettlesome problems. On August 20, 1975, it finally lifted off. Viking 2 followed on September 9. The voyage to Mars would take about 10 months. Each spacecraft weighed 3,527 kilograms, or about 7,775 pounds, when fueled and ready to go. Viking 1 entered orbit around Mars on June 19, 1976, and immediately began taking photographs of possible landing sites. Because of the decision to change the landing site, the Viking 1 lander did not touch down until July 20. A signal from the ground at four in the morning on July 20th ignited several tiny explosive bolts that held the lander and orbiter together, separating the two. Small rockets fired on the lander, easing it out of orbit and into its descent. At an altitude of 19,400 feet, a parachute was supposed to unfurl behind the spacecraft, and at 4,600 feet three rockets would be fired to further slow the lander's descent. Mission controllers at the Jet Propulsion Laboratory waited through an agonizing 19-minute delay—the time it took for radio signals to reach Earth from Mars—for confirmation that the lander was operating normally. "Dignitaries and working engineers alike crowded around the video monitors, listening intently as a calm male voice on the public-address system ticked off the milestones of Viking 1's descent through the atmosphere," Bruce Murray recalled. "'Parachute deployment...Retros firing...TOUCH-DOWN!'" The tension broke as scientists and engineers burst with excitement.

Viking 2 went into Martian orbit on August 7 and dispatched its lander to the surface on September 3. Both spacecraft landed as planned in the northern hemisphere, where it was summertime, the time when any living things were most likely to be active and to reveal themselves. Aboard each lander, squeezed into a one-foot-square compartment, the most elaborate remote-controlled laboratories ever built were ready to go into action. The search for Martian life was set to begin.

When Viking 1 touched down, it immediately sent back a picture of one of the three spidery footpads on which it stood. Mission controllers were relieved to see the footpad resting on top of the Martian surface—not buried deep in dust, as some had feared. The lander then began beaming a series of images back to ground control, giving scientists their first look at the Martian surface. "I remember being transfixed by the first lander image to show the horizon of Mars," Sagan wrote. "This was not an alien world, I thought. I knew places like it in Colorado and Arizona and Nevada. There were rocks and sand drifts and a distant eminence, as natural and unselfconscious as any landscape on Earth. Mars was a *place.* "The first color pictures showed a blue sky, but later examination revealed that the pictures had not been properly processed. The scientists thought the sky would be blue, and so at first they adjusted the color so that the sky was indeed blue—unaware that their preconceived notions were influencing their work. When they re-examined the photos later, they recognized their bias and eliminated it. Then the images revealed a pale, rosy sky—tinted by fine particles of airborne Martian dust.

Viking's measurements of the atmosphere during its descent showed that nitrogen made up 2 to 3 percent of the atmosphere. Nearly all of the rest was carbon dioxide, as earlier observations had shown. Nitrogen is an essential ingredient of life on Earth; its presence in the Martian atmosphere was an encouraging sign. Mariner 9 had estimated that the atmosphere was only about 1 percent nitrogen. But even with a nitrogen content of 3 percent, the Martian atmosphere remained far different from Earth's, which is almost 80 percent nitrogen.

Viking 1 continued to send back images on its first day on Mars, designated "sol 1." (A "sol" is a Martian day. The term is used to avoid confusion with Earth days.) On sol 8, Viking extended its movable arm and dug a small trench in the soil. It picked up a scoop full of Martian soil and dropped it into a hopper from which it would be delivered to the biology instruments. There was an almost immediate response. Murray, then JPL's director, got an urgent call at home the evening after the first data was received from Viking. "Bruce, they've found something," Frank Colella, JPL's press officer, said.

On sol 11, three days after the biology experiments had begun, Harold Klein of NASA's Ames Research Center, the leader of the Viking biology team, appeared at a press conference at JPL, where hundreds of reporters had gathered to cover the story. The gas-exchange experiment designed by Oyama had measured a burst of oxygen two-and-a-half hours after the soil was dropped into the experimental chamber. That was followed by data from Levin's labeled release experiment showing a burst

of radioactive carbon. "The odds were overwhelming that nothing would happen at all," Levin said. "And when we saw that curve go up we all flipped...we knew something was happening there." Horowitz's pyrolytic release experiment had not yet returned any data, but the other two experiments seemed to be showing evidence of life.

Under the glare of television lights, Klein told the assembled reporters that there was "at least preliminary evidence for a very active surface material" and that Levin's results looked "very much like a biological signal." But Klein warned that the results should be viewed with a great deal of caution. Everyone remembered the headlines made around the world a few years earlier when Mariner 7 seemed to find evidence of ammonia and methane in the atmosphere of the planet near the southern polar cap— possible indications of the presence of life. The finding later proved to be the result of an incorrect interpretation of the data. No one on the Viking team wanted to make the same mistake.

Besides, something about the data didn't seem quite right. Oyama thought it would take days or even weeks or months before any organisms would produce enough gases to be recorded in his instrument. Yet the powerful burst of oxygen had occurred almost immediately. Likewise, with Levin's experiment, scientists had expected the amount of radioactive carbon dioxide to grow faster and faster as organisms inside its chamber multiplied. But instead the amount of carbon dioxide seemed to be leveling off after climbing rapidly at first.

Three days later, Horowitz got his first results. His experiment had also produced modest evidence that something was going on, although the data was not what he would have expected. As with the other two experiments, Horowitz's findings were ambiguous. Researchers searched for alternative explanations. The idea was that every other possible explanation should be tested and rejected before concluding that the findings were evidence of life.

An alternative explanation arose almost immediately. This was the surprise that Mars had been saving for the Viking team. The idea was that the surface of Mars was covered with a group of highly chemically active substances of the kind called oxidants. Such chemicals—hydrogen peroxide, used as bleach, is a typical example—react very quickly with water. Oxidants cannot survive in contact with water, but in the parched climate of Mars, such oxidizing agents could be common. The strong ultraviolet radiation bathing the Martian surface could lead to the production of oxidants, the researchers realized. The idea—which had never occurred to anyone before—suddenly seemed highly likely. Oxidants, when dropped into liquid water, would fizz like Alka Seltzer, producing all kinds of unexpected results—some of which would mimic the activity of microbes.

As researchers puzzled over whether they were detecting chemistry or biology, a key piece of data was missing. The GCMS, intended to detect the presence of organic chemicals, had apparently not received its soil sample, and so the experiment hadn't been run. The problem was eventually corrected, and the GCMS went into operation. The instrument failed to find any sign of organic substances on Mars. As Sagan put it, "If there is life on Mars, where are the dead bodies? No organic molecules could be found—no building blocks of proteins and nucleic acids, no simple hydrocarbons, nothing of the stuff of life on Earth."

The experiments were repeated on Viking 2 with similar results. Experiments on Earth demonstrated that oxidants could produce some of the findings seen in the Viking biology experiments. The biology experiments were run until May 1977. After all the effort, after all the testing, the expense and the analysis, the Viking team reported that it could not say for certain whether life existed on Mars. "All experiments yielded results, but these are subject to wide interpretation," the team said in an official report published in the journal *Science* in November 1976. "No conclusions were reached concerning the existence of life on Mars."

Despite the lack of an official conclusion, scientists overwhelmingly accepted the chemical explanation of the findings. Partly that reflects the unwillingness of scientists to accept anything other than the most plausible explanation. To become convinced of the existence of life on Mars, Viking scientists felt they should first rule out every other possible explanation—and in this case, they hadn't done that. The oxidant theory seemed a likely explanation for the findings.

Five years after the Viking mission, Harold Klein, its biology team leader, assessed the mission. "The results of the biological experiments on Viking, taken in isolation, allow the possibility that at least some of the data…might be of biological origin," Klein wrote. But "it would appear more reasonable to ascribe all of the biology 'signals' to nonbiological causation." Perhaps life exists in other regions of Mars, Klein wrote, or perhaps exotic creatures not dependent on carbon remain to be found. But "the available information does not warrant optimism that new approaches will result in different conclusions about biology on Mars." Despite the frustration over the search for life, the Viking mission contributed enormously to knowledge about Mars. The images from the surface alone were enough to justify the project.

Viking marked the close of the first phase of the exploration of Mars. It would be almost 21 years before Martian exploration resumed. The Mars missions, beginning with Mariner 4 in 1964, "transformed Mars in the course of a decade from a mysterious planet with a long, romantic history into one of the most familiar places in the solar system," Horowitz later wrote. The Viking landers were studying the chemical makeup of Martian soil less than two decades after the Soviet Union had made the first space flight, lofting Sputnik 1 into orbit in 1957. Nothing else in space is as fascinating as Mars, and nothing else is likely to generate the kind of effort that went into those first Mars missions. "The like of the Mars project," Horowitz wrote, "with its unique mixture of legend, scientific incentive, technological capabilities, and public enthusiasm, will not be seen again soon.…"

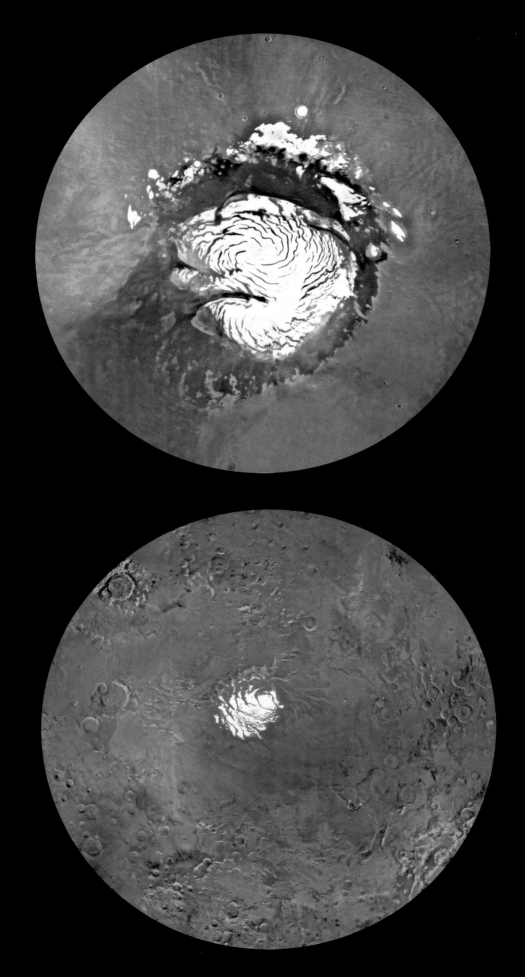

The south polar hemisphere of Mars (lower image) has more heavily cratered terrain and a smaller ice cap compared with the north polar hemisphere.

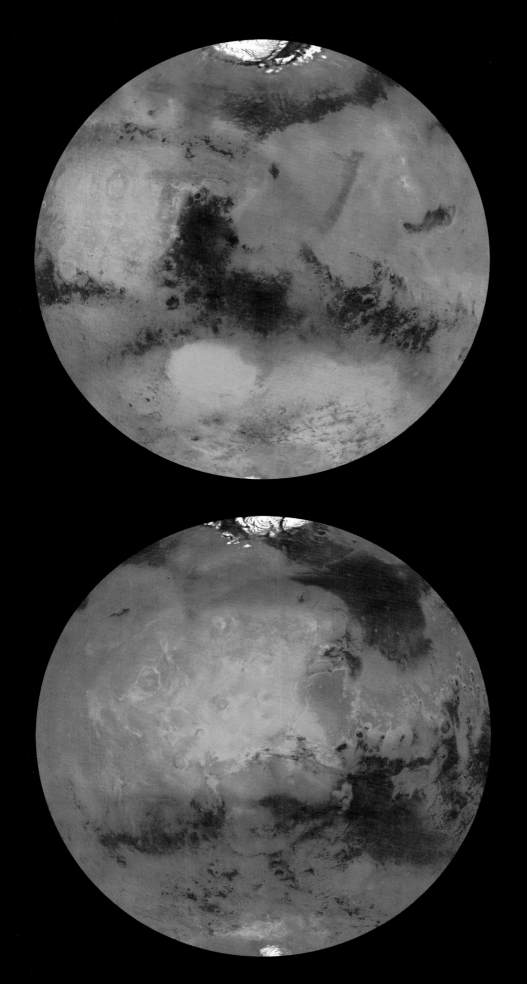

Eastern (upper) and western (lower) equatorial
hemispheres of Mars exhibit dark areas relatively
dust free and perhaps covered with rocks or sand.

V A S T I T A S

—60°

ARCADIA
PLANITIA

Milankovic

ALBA
PATERA

TEMPE
TERRA

ACIDALIA
PLANITIA

CYDONIA
MENSAE

—30°

AMAZONIS

PLANITIA

Olympus
Mons

Ascraeus
Mons

THARSIS MONTES

KASEI VALLES

LUNAE

PLANUM

CHRYSE
+ VIKING 1 LANDER
+ MARS PATHFINDER

PLANITIA

Shalbatana Vallis
Simud Vallis
Tiu Vallis
Ares Vallis

—0°

Pavonis
Mons

NOCTIS
LABYRINTHUS

Arsia
Mons

Mangala Valles

VALLES   MARINERIS

XANTHE   TERRA

SINAI
PLANUM

MARGARITIFER
TERRA

DAEDALIA
PLANUM

SOLIS
PLANUM

—30°

TERRA

ICARIA
PLANUM

AONIA

ARGYRE

PLANITIA

Galle

Copernicus

Lowell

TERRA

SIRENUM

—60°

Phillips

Schmidt

The surface area of Mars, roughly equivalent to the
area of the continents on Earth, has heavily
cratered terrain in the south, including two giant
impact basins (Hellas and Argyre) that probably
formed roughly four billion years ago. The lightly
cratered northern lowlands include Utopia,

Acidalia, and Arcadia; and Isidis, Elysium, and
Chryse Planitia. The Tharsis province includes the
four giant volcanoes (the three Tharsis Montes and
Olympus Mons), Alba Patera, and Valles Marineris
and the surrounding areas. Elysium is a smaller
volcanic area in the eastern hemisphere. Large

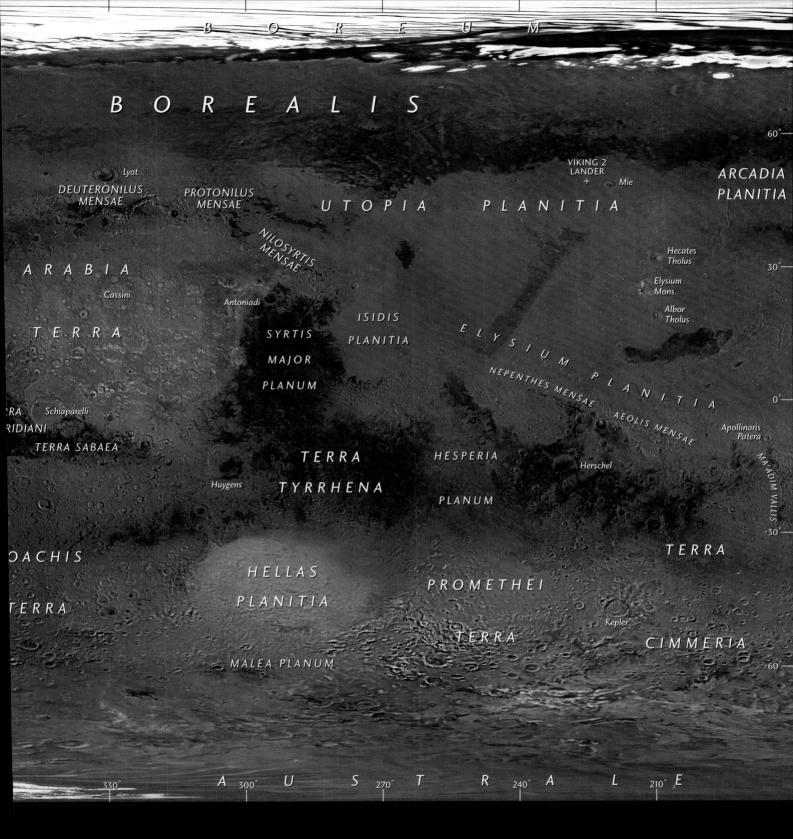

B O R E U M

B O R E A L I S

60°

Lyot

DEUTERONILUS
MENSAE

PROTONILUS
MENSAE

VIKING 2
LANDER
+   Mie

ARCADIA
PLANITIA

UTOPIA   PLANITIA

NILOSYRTIS
MENSAE

Hecates
Tholus

30°

A R A B I A

Cassini

Antoniadi

Elysium
Mons

Albor
Tholus

T E R R A

SYRTIS

ISIDIS

PLANITIA

E L Y S I U M   P L A N I T I A

0°

MAJOR

PLANUM

NEPENTHES MENSAE

AEOLIS MENSAE

Apollinaris
Patera

RA   Schiaparelli

RIDIANI

TERRA SABAEA

T E R R A
T Y R R H E N A

Huygens

HESPERIA

PLANUM

Herschel

MA'ADIM VALLIS

-30°

OACHIS

HELLAS

PLANITIA

PROMETHEI

T E R R A

CIMMERIA

TERRA

MALEA PLANUM

T E R R A

Kepler

-60°

A   U   S   T   R   A   L   E
330°   300°   270°   240°   210°

channels drain from the highlands into Chryse
Planitia. Both Ares and Tiu Valles perhaps flooded
the Pathfinder landing site. The Kasei Vallis floods
may have affected the Viking 1 landing site. Viking
2, in Utopia Planitia, lies just west of the crater
Mie. The larger, northern polar cap includes both

water and carbon dioxide ice. The smaller, southern polar cap, consisting mostly of carbon dioxide,
changes size dramatically with the seasons. The
extremely warm, long southern summers (and
long, frigid northern winters) on Mars result from
the tilt of the planet and its highly elliptical orbit.

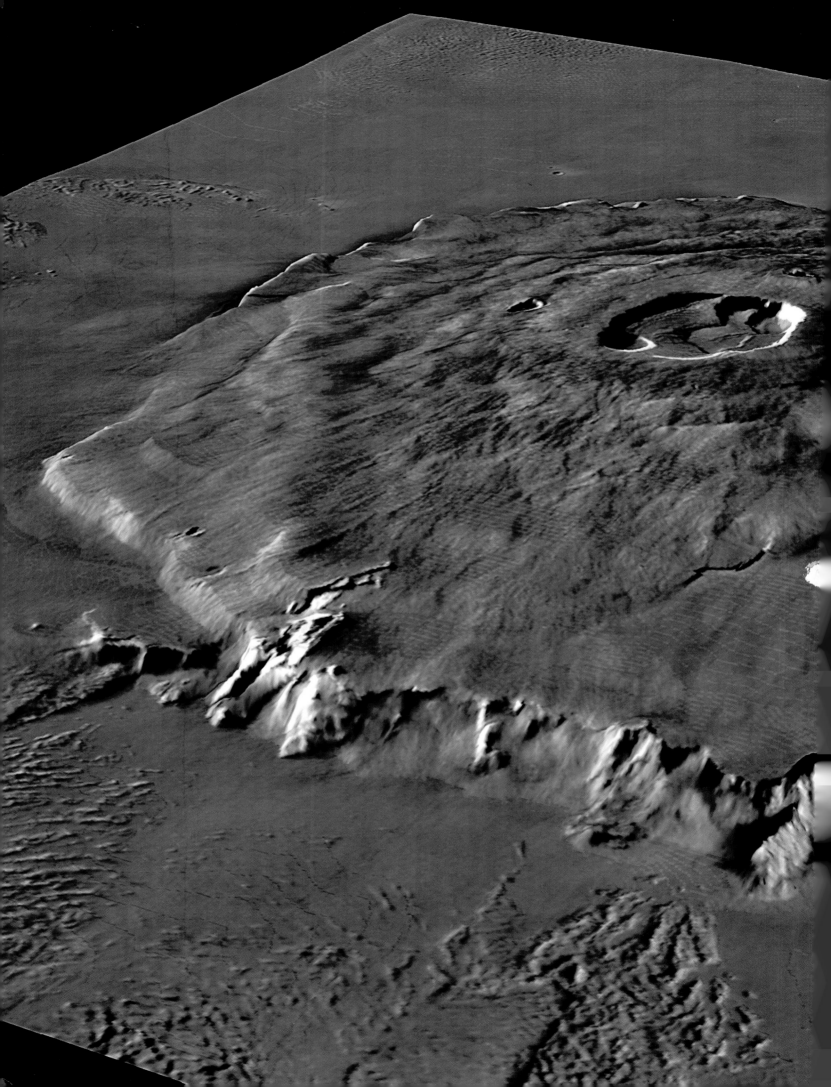

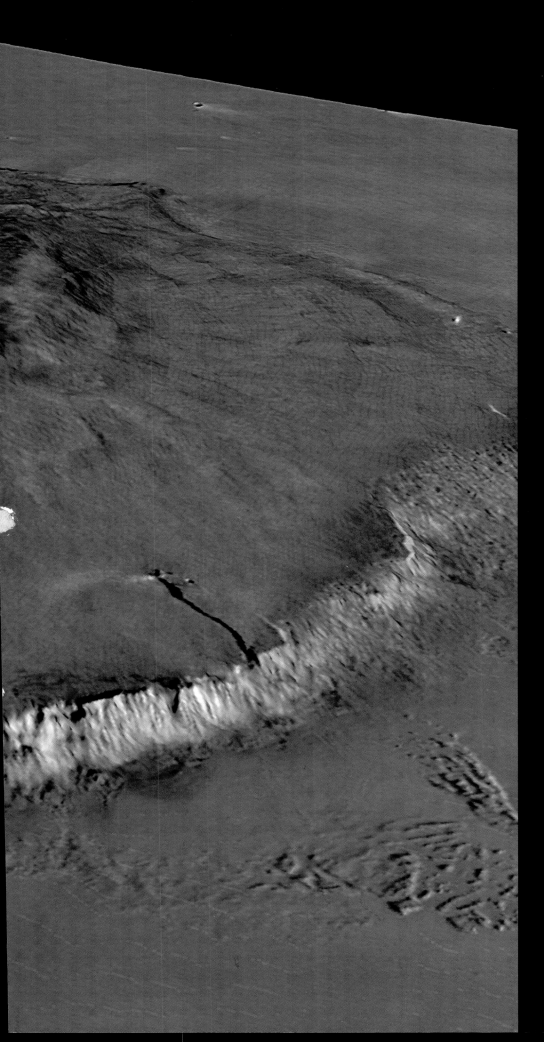

This artist's rendering depicts an oblique view of Olympus Mons, looking toward the southwest. The central caldera on top gives way to shallow slopes typical of this type of volcano and lava. Including the huge escarpment at its base—up to 6 miles (10 km) high—the volcano rises vertically almost 17 miles (27 km). Its base would cover the state of Arizona. The slopes and morphology of Olympus Mons point up its similarity to Hawaiian volcanoes, composed of basalts produced by melting at the top of the mantle. Olympus Mons's few impact craters indicate activity in the recent geological past.

FOLLOWING PAGES: This cylindrical projection of Mars uses color to represent elevation. Four giant volcanoes, including Olympus Mons, show as reddish orange; the Hellas basin, as blue. The huge, high-standing Tharsis Bulge, seen here as yellow-green, sits astride the highland-lowland boundary.

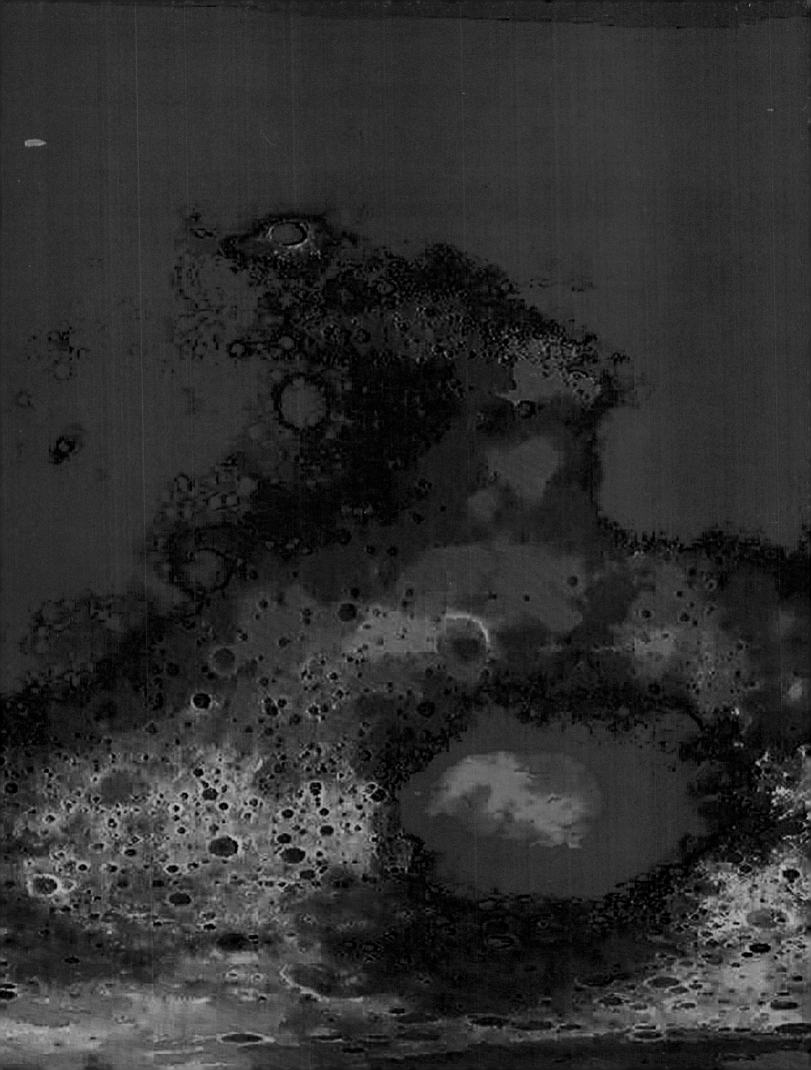

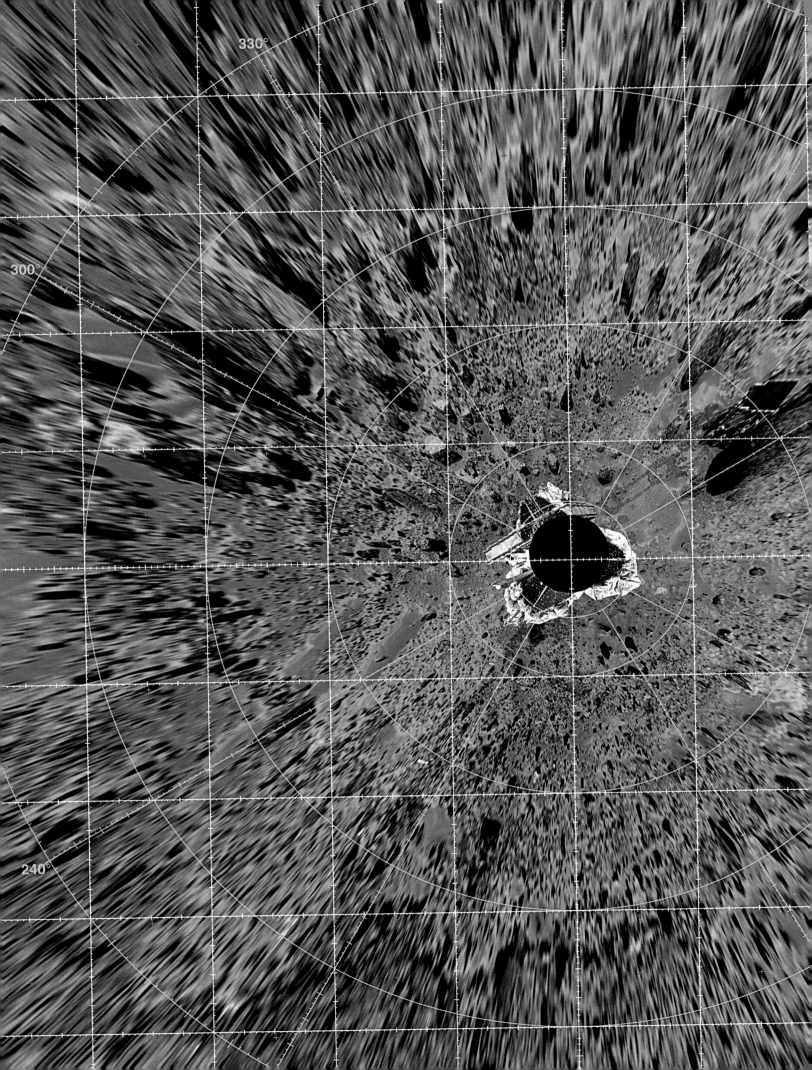

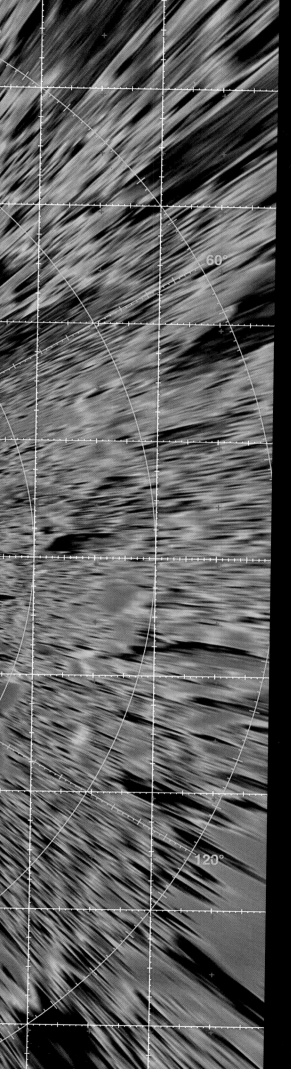

CHAPTER FOUR

PATHFINDER: A NEW ERA IN MARS EXPLORATION

Viking planetary exploration entered a period that Bruce Murray has described as "lost in space" for two decades. NASA became absorbed by the development of the space shuttle, which was supposed to be a workhorse vehicle for carrying satellites into orbit. The shuttle, while capable of doing some science, was not primarily intended to conduct research. In 1977, two Voyager spacecraft were launched on a tour of the planets, and they sent

Center of map-view projection shows the Pathfinder lander with its petals, air bags, and rover ramps. The rover is next to Yogi, making a chemical analysis measurement.

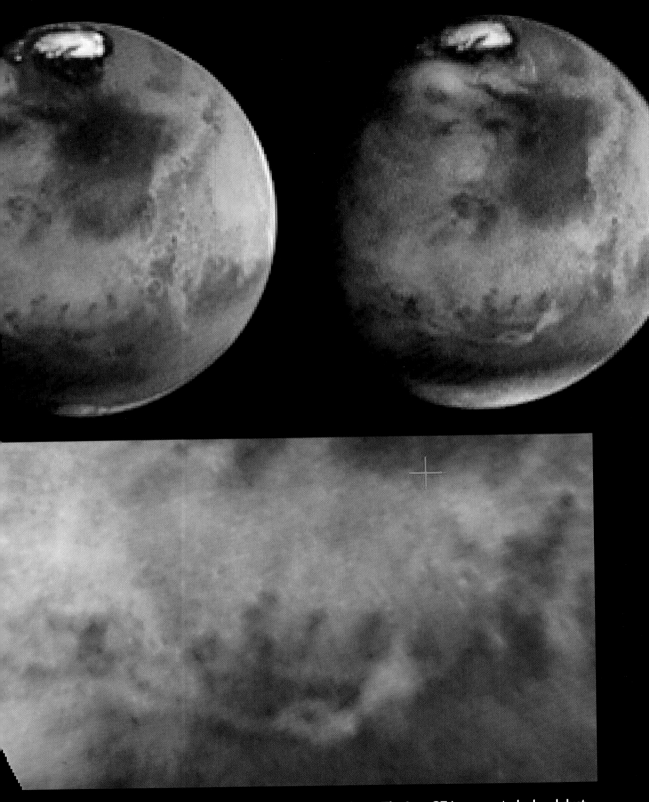

Hubble Space Telescope imaged the western hemisphere of Mars (with north polar cap) on May 17 (left) and June 27, 1997 (right and lower), before the Pathfinder landing. The May image showed some white clouds and a clear, cold atmosphere. The June 27 image noted a local dust storm in the canyons of Valles Marineris, diffusing north toward the landing site (green cross). Pathfinder scientists expected a clear or slightly dusty atmosphere for landing on July 4, 1997.

photographs back to Earth in 1979 and during the 1980s. But the space shuttle moved NASA's focus away from planetary exploration. Following the Voyager launches, the Jet Propulsion Laboratory devised a mission to Jupiter as the next logical step for a continuing program of planetary exploration. Unlike Voyager, which would fly by Jupiter, the JPL mission would orbit Jupiter and send a probe into its swirling, gaseous surface. Narrowly averting cancellation of the mission by a Congress now indifferent to NASA and skeptical of expensive space missions, the space agency informed JPL that the probe would be launched on the space shuttle. Murray, who was the director of JPL at the time, was concerned about the decision. Because of the relative positions of the planets, the deadline for launching the Jupiter mission was January 1982. If the launch opportunity was missed, the mission would be lost. John Yardley, head of the space shuttle program, assured Murray that the shuttle would be ready. "Hell, Bruce," Yardley said, "we have 18 months of pad." That is, an extra 18 months had been built into the schedule. Even if the shuttle program fell as much as a year and a half behind, it would still be completed in time to launch the Jupiter mission in January 1982.

In March 1979, Voyager 1 began sending back spectacular photographs of Jupiter and its moons. On July 9, Voyager 2 arrived, sending back more photographs. Although Voyager couldn't compare with Viking's landing on Mars, it rekindled public enthusiasm for space science. On July 20, 1979, the third anniversary of Viking's landing on Mars, Murray received a phone call at JPL from Yardley. The shuttle, Yardley conceded, had fallen so far behind schedule that it would not be ready in time to launch the Jupiter orbiter. "NASA's obsession (with the space shuttle) had claimed its first victim," Murray said.

The space shuttle consumed NASA's energy and, more importantly, its budget during the 1980s. The first sign of change came in 1989, on the 20th anniversary of the moon landing. President George Bush chose that occasion to make a speech calling on NASA to begin planning a manned mission to Mars. Standing on the steps of the National Air and Space Museum, Bush said it was time for the United States to begin "the permanent settlement of space," to go "back to the Moon, back to the future, and this time back to stay....And then, a journey into tomorrow, a journey to another planet: a manned mission to Mars." Some said that would require a doubling of NASA's then 13.3-billion-dollar-budget. Estimates of the cost of going to the moon and Mars ranged from 400 to 800 billion dollars.

NASA began developing a long-range plan that could ultimately lead to a manned mission to Mars. First, though, NASA wanted to lay the scientific and engineering groundwork for such a mission. In 1992, the agency announced a plan to send a series of small, low-cost spacecraft to Mars to demonstrate that it could dramatically reduce the cost of space exploration. The first of the low-cost flights was named Pathfinder.

This was the beginning of a new NASA, guided by the agency's new watchwords: "Faster, better, cheaper." NASA was going to reinvent planetary exploration. In the early 1990s, a plan was devised to dispatch anywhere from 12 to 20 tiny landers to Mars. The project was called the Mars Environmental Survey, or MESUR. The MESUR landers would take simple weather and other measurements around Mars, providing a new set of data that would encompass the entire planet. At about the same time, NASA also began talking about what it called Discovery missions—small, inexpensive planetary spacecraft that would be built for about 150 million dollars each. To put that figure in context, the Viking mission to Mars would cost an almost unimaginable 3 billion dollars if flown again in the 1990s. Somehow, the Mars environmental project and the plan for Discovery missions crossed, and NASA decided to launch a demonstration flight. The plan was to show that the Mars project was feasible and to show that a spacecraft could indeed be built for 150 million dollars. The mission was Pathfinder.

Originally designed to demonstrate new spacecraft technology, Pathfinder was later equipped with a tiny rover able to conduct scientific analyses of Martian rocks. The rover added an important scientific component to Pathfinder's mission, along with an element of drama and excitement that would later captivate the public.

Work on Pathfinder began at the Jet Propulsion Laboratory in early 1992. Engineers were thrilled with the prospect of once again landing on another world. There was only one problem: Nobody remembered how to do it. No attempt to land on another planet had been made since Viking, and many of the people responsible for that mission had since left JPL. Mission planners began assembling a team of scientists and engineers, drawing on expertise from all over the lab. An early member of the team was Robert Manning, a JPL engineer who would become chief engineer on the Pathfinder project. He was 34 years old. Manning later assumed the leadership of the team that would plan what was known as Pathfinder's "EDL"—its entry, descent, and landing. Manning had just graduated from high school in 1976 when Viking landed on Mars. He remembered the mission vividly from the perspective of a young space buff, but he certainly had no knowledge of the technical engineering details that made Viking's landing possible.

So he did the obvious thing: he began searching through the corridors of JPL for the people who could give him the background, who could explain to him how it was done before, and who could help prepare him to do it again. He soon began to see that it wouldn't be that simple. "At JPL, we know how to build spacecraft," Manning said. "We've built spacecraft for a long time. But they are mostly spacecraft that cruise between the planets or go into orbit around planets. We haven't built landers in a long time." JPL's expertise in landing proved to be almost nonexistent: Nearly everyone who had worked on Viking had retired, and some had died. Manning searched elsewhere in NASA, attempting to locate scientists or engineers who might be able to help, and again he was disappointed. "There is no institution anywhere in NASA that specializes in landing systems," he said. "There was no organization you could go to and say, 'How did we do this then, what was going on?' We had to create out of whole cloth—including going back to retirees who worked on Viking." He found a small private company that had done some research on landing space vehicles on Earth, but that was little help. "The problems in landing on Earth are quite different from Mars—the atmosphere is vastly different," Manning said.

With little guidance, Manning began to think about the problem. JPL was well equipped to design a spacecraft, launch it, get it to Mars and put it into orbit. He began to visualize the landing as the equivalent of placing the spacecraft into an extremely low orbit—so low, indeed, that it ultimately touches the ground. Manning realized that the task would be extremely complicated. "The process of getting to the surface was viewed as a mechanical engineering problem. But the mechanical engineering was only part of it. There was navigation, there were the issues associated with aerodynamics, the software, the electronics, the power system—all these pieces had to come together. The entry, descent, and landing was a property of this complex spacecraft, and there was nobody focusing on it in a systems way, pulling it all together." Even if Manning could find information on the Viking landings, it would be helpful only to a point. Viking dispatched its landers from Mars orbit; Pathfinder was going to land on a direct shot from Earth. "It wasn't just that we solved the problem again with this spacecraft—we'd never solved these problems before," Manning said. "A lot of spacecraft have problems that are unique—but we had them in spades. And with a low budget, we really had to pull out all the stops."

Manning found the problem of landing a spacecraft on Mars so intriguing that he decided to assign it to himself, and mission managers agreed. The move was typical of what would come to be one of Pathfinder's hallmarks: a fluid, ad-hoc management style designed to break out of the old bureaucratic mold and simply assign tasks to those who seemed best able to complete them, whether or not the decisions made sense from the standpoint of an organizational chart. The Pathfinder team had been assembled specifically for this project. Members were chosen for their individual talents and for their ability to work with others, not because they had standing or asked for the job. Some who joined the team were later let go. Many of those who left the team were talented, hard-working scientists and engineers, but they weren't able to handle the project's unusual demands or were temperamentally unsuited to Pathfinder's freewheeling approach. "We had a fairly small, super-talented team without a lot of bureaucracy," Manning said. "We changed our management scheme to fit people's talents." Mission directors had a budget of $150 million, excluding the costs of the rover, launch vehicle, and spacecraft operation. They felt they had little choice but to manage the project the way they needed to. "You could go down the path with the traditional management approaches and find you're out of money or out of time. We needed an incredible diversity of talent and experience, and we organized ourselves around solving problems." That helped lay the groundwork for what would become an extraordinary group. "Initially very few people knew each other," Manning said. "It certainly wasn't true at the end. We were all an intense family when Pathfinder ended."

Unable to find many engineers at JPL who could help him, Manning began to search for the documentation of the Viking mission. One might guess that in a high-profile, billion-dollar planetary exploration mission such as Viking, engineers would leave behind endless reports, file drawers full of bulging notebooks, and reams of scientific data. Not so, Manning discovered. Viking's principal scientific findings had been published in science journals, but many other scientific and engineering details had never been recorded. Viking scientists had produced computer models of the Martian atmosphere, for example, revealing wind and weather patterns that Manning needed to understand in order to plan

Pathfinder's parachute landing. But before Manning could use the data from the models, he needed to know how they were put together—and that was the tricky part. "We had to peel back the onion to figure out what assumptions were made, how the data would apply to Ares Vallis, our landing site, and how it applied to time of day," Manning recalled.

A more important problem was Pathfinder's parachute and aeroshell—the casing that would shield it from the heat of entry into the Martian atmosphere. Pathfinder, like Viking, would use a parachute to slow its descent toward the Martian surface. Indeed, Pathfinder inherited Viking's parachute—but without knowing for sure whether it would work. Manning thought that Pathfinder, with its beeline track from Earth to the Martian surface, bypassing orbit, might require modifications of the parachute. And Pathfinder had an entirely different landing system. The Viking landers were slowed by retro-rockets until they gently touched down. Pathfinder was wrapped in air bags and designed to bounce like a giant beach ball along the Martian surface until it came to rest. "The choices that the Viking people made—why it was the way it was—weren't clear until you really studied it," Manning said. He concluded that "if you were designing from scratch, you would never use their parachute or their aeroshell for Pathfinder."

The aeroshell, for example, was cone-shaped, designed for a vehicle that could control its position. "When Viking hit the atmosphere, there were thrusters that kept it properly oriented. It actually tried to fly to its landing site. Pathfinder is just a rotating bullet with nothing controlling it. This cone shape produces some unstable results—not so unstable that it's devastating, but you live with that."

Viking used its parachute to slow down—nothing more. On Pathfinder, however, the parachute had to do something else—it had to hang straight up-and-down, with as little back-and-forth wobble as possible. This positioning would point the rockets on Pathfinder directly at the ground, so they would slow its fall. If it bobbed like a World War II paratrooper swinging wildly below his parachute, Pathfinder

A six-lobed inflated air bag surrounds each face of the tetrahedral lander to cushion it while landing. The air bags had internal bladders covered by layers of abrasion-resistant material to prevent them from tearing on sharp-edged rocks. The air bags worked well, with the lander bouncing over 15 times in one of Mars's rockiest regions without puncturing.

could be in trouble. "If it's tilted when the rockets fire, they will do a good job of stopping it, but they will also kick it sideways. You've got this 17-foot beach ball that could take off horizontally at a pretty decent clip, up to 60 miles an hour," Manning said. That meant the beach ball, rather than bouncing more-or-less up and down, would go rolling and bouncing across the surface "like a Michelin tire," Manning said. This is a beach ball that weighs 1,973 pounds at launch (fueled)—when it hits the ground, it hits the ground hard. "Once you throw the Michelin tire job in on top of this normal impact job, you realize you've got a really hard problem."

Data from Viking showed that some of the rocks on the Martian surface were sharp and some, in effect, were bolted down to the surface. "Some of them are bonded, with a sharp pinnacle and the rest of the rock in the sand—not going anywhere. Imagine all these rocks on a runway. This ball coming at a shallow grazing angle would spin like a tire on a horrible nasty-looking terrain." The danger was that the fabric the air bags were made of could rupture, allowing a suddenly unprotected Pathfinder to be dashed to pieces.

Manning decided that the parachute would have to be modified to make it more stable. But that created a problem: Several of the few remaining Viking alumni who could be located were now on Pathfinder's review board. It was their job to report to NASA headquarters in Washington whether this young, untested team of Pathfinder engineers was doing a good job. "They were skeptical, incredibly skeptical—borderline hostile," Manning said, "as they should be. They were paid to challenge everything. So it was a big deal when we deviated from the Viking heritage....It takes a lot of work to explain to a Viking manager why we decided to dump his parachute, a perfectly good hundred-million-dollar project." The Viking alumni on the review board wanted to know if Manning planned to spend hundred million dollars himself to do an equally thorough job of designing a new parachute.

Of course, he couldn't. What Viking had spent on its parachute was two-thirds of Pathfinder's entire budget. Manning had five million dollars to develop Pathfinder's parachute. He discovered, however, that Viking's designers had studied the stability of their parachute. The Viking engineers had themselves proved that changing the shape of the parachute would make it more stable. That data was discovered accidentally by a member of Manning's team who had been scanning articles on Viking in scientific journals. After the changes were made, Pathfinder's parachute resembled Viking's, but Pathfinder's had a thicker "band," the ring of fabric around the bottom of the parachute, underneath the rounded top.

The solution of the parachute problem was a study in what Pathfinder engineers came to call "unknown unknowns." In JPL terms, the parachute problem "was an unknown unknown that became known," Manning said. What he meant was that the size of the problem gradually became clear as the engineers worked on it. While some were busy studying parachute stability, for example, others were solving another knotty problem—how to test the beach ball's landing capability. As each group progressed, it gathered information that would help the other group.

The test group decided to make use of a huge, 12-story vacuum chamber at a NASA facility near Cleveland, to try to duplicate conditions on Mars. Inside the chamber, Manning and his team built a "Martian surface" out of wooden platforms, steel girders, and lava rocks. "We bolted on lava rocks shipped by train from California, because you can't get lava rocks in Cleveland," Manning said. Then they rigged up cables and bungee cords to drag the air-bag across the rocks.

"Our first tests were total disasters," Manning said. "The first time we did it, we had a tear the size of a human being—a six-foot-tall hole torn into the fabric. In the high speed video tapes, you could see right through the hole all the way to the lander." Meanwhile, the results of other tests going on simultaneously were turning the "unknown unknown" into a known unknown. The redesigned parachute appeared to be more stable than engineers thought it would be—meaning the beach ball wouldn't have to endure as much wear-and-tear skidding along the surface as initially feared. And additional study of the aeroshield showed that it could withstand more weight than originally thought—so Manning was able to add a few more layers to the airbag. Manning's

entry-descent-and-landing team was dodging bullets from every side. "There is a scene in *Star Wars* where the *Millennium Falcon* is zipping through an asteroid field and just missing all these asteroids and making it through with incredible luck," he said. "I felt that way the whole time." By this time, it was only one year until launch. If unanticipated problems had developed, they might have doomed the mission.

In 1995, in fact, the reviewers overseeing Pathfinder—a group of NASA scientists and engineers, including a few veterans of the Viking missions—came close to canceling the Pathfinder mission. "They felt the risks were so high, with all these technical things going wrong, that it was worth canceling," Manning said. In late 1995, steps were taken to begin formal consideration of canceling the mission. The Pathfinder team at JPL argued vigorously against it. "We said, 'We're getting there now.' Our nervousness factor was dropping quickly; we could see light at the end of the tunnel." What the overseers viewed as a series of problems, the scrappy Pathfinder team viewed as a series of solutions.

The Pathfinder team was going to need a bit more luck, though, before it was through. In December, 1995, Manning and the entry-descent-and-landing team took a mock-up of Pathfinder's heat shield to China Lake Naval Weapons Center, about 150 miles northeast of Los Angeles, for a test of Pathfinder's retro-rockets. The rockets seemed to fire beautifully, but the computer monitoring the test crashed into the ground, so there was no confirmation that everything had gone according to plan. Manning wasn't concerned. "We knew the test had worked properly....At least, it sure looked that way." The Pathfinder engineers went back to JPL, and most of the weapons center personnel left for Christmas vacation. When they returned in January, they found the test computer buried in the desert sand and discovered that it was intact, although its battery was almost dead. The data was downloaded and analyzed just before the battery died—and Manning had a new problem.

The data from the computer revealed that the rockets had stopped firing prematurely—something that wasn't apparent to Manning and others who watched the test from the ground. The chance retrieval of the computer data had turned what appeared to be a successful test into "something very, very scary," Manning said. The data showed that shortly after the rockets started firing, they began to vibrate and all three self-destructed. If that were to happen on Mars, it could lead to a crash landing that would destroy Pathfinder. "This was January 1996 and we were launching at the end of the year. It scared us out of our wits," Manning said. He immediately sought help from rocket experts, but couldn't find any. "You talk about 'rocket scientists'—well, there aren't any. They're dead—or retired." Even Thiokol Corporation, the builder of the rockets, had few people left who understood what happens when rockets fire.

After talking to the few people he could find, Manning concluded that the vibrations from the three rockets were reinforcing one another and ripping the rockets to pieces. And the source of the problem was determined. At the request of Pathfinder's scientists, Manning had changed the formula for the rocket fuel. He had removed alumina, which the scientists feared could contaminate the landing site and interfere with their geological testing. No one remembered why the alumina was there until the vibration problem appeared at China Lake. Alumina, Thiokol said, was added to reduce vibration as the fuel burned. Manning suggested that the alumina could be put back, because Pathfinder was likely to bounce away from the point where the rockets would be fired—so the alumina would not contaminate the site where the lander finally came to rest. The decision was made to restore the alumina. Another test was run at China Lake, and the rockets performed perfectly. By the time the problem was solved, it was June 1996—less than six months before launch. "These are the kinds of problems we had, some of the more exciting ones," Manning said. As each problem appeared, it was solved, often barely in time to keep on schedule. But for Manning, the fear never left. "We were afraid there would be problems with no solutions, and we just wouldn't get there."

The idea of putting a small, roving vehicle on Mars was raised during the 1980s. When Rob Manning began thinking about how to put Pathfinder safely on the Martian surface, the spacecraft did not include the tiny rover that would turn out to be one of the highlights of the mission. "Pathfinder was unique because it was trying to be done so cheaply," Donna Shirley, manager of NASA's Mars Exploration Program, said a few months after the mission was successfully concluded. "And trying to run a network of missions cheaply was dependent on being able to land cheaply. So Pathfinder was set up to see if we could learn to land on Mars cheaply." The idea of using air bags to convert the spacecraft into, in effect, a giant beach ball arose early, partly in reaction to Viking's use of retro-rockets. "The airbags were believed to be a cheap way to land, mainly because Viking used retro-rockets and Viking was expensive—ergo retro-rockets are expensive. Which turns out not to be true," Shirley said.

Although Americans hadn't tried the idea before, the Russians had. "They had landed successfully on the moon and on Venus, and we were told that they did that using parachutes and airbags. In fact they claimed that they landed on the moon with just the air bags, though there seems to be some skepticism about whether that really worked or not." The Russians' idea was to build the lander in the shape of a sphere. Once it has bounced to the ground, the airbags are deflated and the sphere opens like a flower, lowering its petals to the ground. The problem with that idea is that the petals are curved and not ideal as a platform for scientific instruments. Pathfinder engineers decided to use the concept, but they changed the design to a tetrahedron—a small, four-sided shape. When the spacecraft landed, the three upstanding sides would fall open, flowerlike. Not only would the flat panels be better for mounting instruments, they were expected to be cheaper to manufacture. "In this country, we don't do spherical spacecraft," Shirley said. "That's what the Russians do."

When the Russians had used air bags, they had allowed their spacecraft to bounce repeatedly until they finally came to rest. The air bags were then deflated. The strategy required that the spacecraft be built tough enough to withstand all those bounces. JPL engineers decided to take a different approach. They would put explosive plugs in the air bags and blow them out just as the spacecraft landed. "Instead of a superball, they wanted something like a nerf ball—it just squishes," Shirley said. It's analogous to what happens in a car crash—the fenders and front of the car collapse, absorbing energy that would otherwise be transmitted to the driver. "This was the same idea. You would have an air bag that would let the air out as it landed."

The engineers ran some tests, and they discovered that their version of the air bag system would produce a spectacular crash. "It turned out that if you just let the air bag plugs blow out when Pathfinder hit, they wouldn't deflate fast enough, and you'd still bounce." The next time Pathfinder came down, the airbags would be deflated—and the spacecraft would be destroyed. Next, the engineers tried putting timers on the plugs, so they could be blown out just before Pathfinder hit. But the engineers couldn't determine exactly when Pathfinder would hit, so that plan was scrapped. In the end, they dropped the nerf-ball idea and went back to the bouncing beach ball. They added inexpensive military retro-rockets that would be fired while Pathfinder was descending to slow it down. And they decided to let the spacecraft bounce and come to rest before deflating the air bags.

Another innovation with Pathfinder was its computer. Previous spacecraft had always used custom computers built at JPL that could satisfy the unique demands of spaceflight and also withstand the radiation and the extremes of hot and cold present in outer space. When Pathfinder was being designed, IBM offered JPL a discount price on a production-model RS 6000 computer. IBM hardened it against the ill effects of radiation and asked JPL, in return, to do the testing and modifications required to qualify the RS 6000 for spaceflight. The idea was that IBM would then have a product it could sell for future spacecraft. JPL took the deal, wrapping the RS 6000

chip with memory and special components to handle all the inputs and outputs on the spacecraft. It was equipped with power converters to allow it to run off of batteries and solar power. "There was a lot of development around this little chip," Shirley said. The computer with its software ultimately cost about $12 million.

The development of Pathfinder continued, but it was still without its rover. JPL had been working on rovers since the 1960s. It had explored vehicles with legs, wheels, tracks, and anything else it could think of. The Soviet space program had also experimented with rovers on satellites, and it had actually sent satellite-borne rovers to the moon. The Russians had devised a rover on wheels that was capable of climbing stairs, but it wasn't particularly adept at dealing with obstacles like the rocks found on the surface of Mars. JPL came up with a new design, taking its inspiration from wheels on a train. They are mounted in such a way that they can move up and down to allow for roughness in the tracks. JPL engineers designed a system in which the wheels were mounted on levers, so they could move up and down to accommodate roughness on the Martian terrain without transferring all of that motion to the rover itself. That gave the rover advantages over previous designs, including the Soviet design. The rover was able to climb over obstacles one-and-a-half times the size of its wheels. If the wheels moved up to clear an obstacle, the rover body would tip only half as much. That was important for providing a stable platform for scientific instruments. Also, because the top was covered with solar cells to provide power, Pathfinder's engineers wanted it to remain relatively flat. If it tipped too far away from the Sun, it would lose power.

Designing the rover was only part of the problem. The next question was: How do engineers on the ground drive it? The round-trip light time to Mars—the time it takes for mission controllers to see what's happening on Mars and to get a command back there—can be as long as 20 or 30 minutes. "If you're looking through the rover's eyes, and you see a cliff coming and you say 'Stop!' it's too late—it will be over

Air bag-retraction and petal-opening tests perfected the lander's opening into operational configuration. In upper photo, the lander has determined which face points downward, has retracted the air bags of the other petals, and has begun to open the petal on the ground to right itself. In the lower photograph, the lander has almost completely opened the remaining petals.

the cliff," Shirley said. "So it has to be smart enough to stay out of trouble."

For an earlier mission that never got off the ground, JPL had experimented with rovers equipped with artificial intelligence—that is, with computers sophisticated enough to analyze what lies before them, pick out the hazards, and direct a rover around them. Another idea was to take pictures from orbit that would show every obstacle bigger than one meter in size, and to design the rover to be able to cope with one-meter obstacles. "We built a rover that had one-meter tires, and it was bigger than a pickup truck," Shirley recalled. "Some of the original rovers were too small and their computers were too big, in that day, so they would drive around dragging cables that ran back to the lab."

Engineers then devised a little eight-inch-long rover called Tooth. Scurrying around on its four wheels, it had enough intelligence to carry out a few simple tasks. "You could put a light in the middle of the room, and it would run away from the light," Shirley said. "Then it would look for what looked like hockey pucks—little flat, black disks. When it found one, it would clamp on to it and then turn around and go toward the light. It would put the disk down at the bottom of the light and then run away from the light again." It could also follow walls and carry out a few other simple commands. Engineers put Tooth's brain on a slightly larger rover called Rocky that had been under development for some time, creating Rocky 3. "They took that out in the desert and ran it and it had great mobility and could actually do things. They staged a demonstration where they built a mock lander with a hopper. Rocky 3 went out and collected some soil and came back and dropped it in the hopper—all autonomously." This was in the early 1990s, and for the first time, after years of development, the rover team finally had something that might be ready to go into space. Rocky 3 weighed about 50 pounds, which made it too big for one of the small Mars environment landers then being discussed.

The Mars environment landers had not won universal acclaim, however. Meteorologists liked them, because they would provide good data on Martian weather. But the geologists were not happy. "Geologists need mobility to get up close to rocks and to put instruments on them. Viking's experience had told us that just reaching out with an arm and trying to do something with rocks didn't work. So we were seeing that mobility was the answer—but it had to be small mobility," Shirley said. Money for rover development was in short supply, but engineers scraped together about two-and-a-half million dollars to try to miniaturize Rocky 3.

An engineer by the name of Lonnie Lane, who had built some of the instruments for the Voyager spacecraft and "is famous for doing wild and crazy things," according to Shirley, shrank the rover from 50 pounds to about 15 pounds. He put a camera on it and a newly developed seismometer about the size of a pack of cigarettes. In June 1992 JPL engineers took the new rover—Rocky 4—out to the desert and showed that they could drive it around, chip a sample off a rock and put test instruments up against rocks. It was now time to fly it.

Work on Pathfinder had begun and was directed primarily by engineers trying to demonstrate the feasibility of safe, inexpensive landings on Mars. But with the rover, NASA's scientists had the opportunity to add some science experiments to Pathfinder. As long as NASA was going to Mars, why not try to piggyback a bit of scientific research onto the engineering mission? German space scientists offered to provide an instrument called an alpha-proton x-ray spectrometer, which could be used to identify the elements making up different rocks on the Martian surface. Soon scientists were swarming over this engineering mission, drawing a protest from Tony Spear, Pathfinder's project manager. He didn't see how Pathfinder could incorporate the scientific instruments and still remain within its 150-million-dollar budget.

Finally a compromise was reached. Pathfinder would take along some of the instruments the scientists were proposing, but not all. It would have a camera and a weather station. And, in a victory for geologists, it would accept the German offer of the alpha-proton x-ray spectrometer. "The only way to get the alpha-proton x-ray spectrometer to a rock, which is what the geologists wanted, was to have mobility," Shirley said. "The engineers looked at arms, and they looked at bouncing the instrument out with a spring. And the only way they could guarantee to get it out there was to have a rover. And here was this free rover." If Pathfinder took along a rover, the geologists would get their science, and the rover team would get a chance to take its project into space for the first time. The rover development program had collided with the Martian exploration program, and Pathfinder now had a rover. The first of NASA's "faster, better, cheaper" missions was ready to go.

As planning for Pathfinder began, though, an example of old-style planetary exploration remained in the pipeline. In 1992, NASA launched a 900-million-dollar mission called Mars Observer, with a 5,700-pound spacecraft that would mark America's first return to Mars in 17 years. Launched on September 25, 1992, it would arrive at Mars in August 1993. Mars Observer was an orbiter designed to scan the entire Martian surface and send back images with much finer detail than any before taken of Mars. It was expected to send back more information on Mars than all previous spacecraft combined. It carried a neutron detector that could search for water in the top meter of Martian soil and also look for carbon-containing materials. Alas, none of that data was ever collected. The spacecraft performed reasonably well until it reached Mars. Just before it was to go into orbit, contact was suddenly lost. Nobody knows exactly what happened. The leading theory is that some of the spacecraft's fuel leaked through a valve, froze, and formed a solid plug in the fuel line during the long, cold trip to Mars. When the valves were opened to ignite the engine and put Mars Observer into orbit around Mars, the theory goes, the fuel line ruptured, sending the spacecraft spinning out of control. "That was the first planetary mission we had lost in 27 years," Shirley said. "We had lost individual spacecraft, like Mariner 8, which went in the drink because the launch vehi-

cle didn't work. But we had never lost whole missions before—because we used to fly two of everything. Now we were only launching one."

Mars Observer proved to be the last of the massive, old-style Martian space probes. The return to Mars now focused on Pathfinder. The Mars Observer spacecraft had cost 400 million dollars, and the cost of the mission, including launch and ground operations, had approached one billion dollar. The entire Pathfinder mission, in contrast, was budgeted at about 250 million dollars—a quarter the cost of the Mars Observer mission. Pathfinder marked a radical revision in NASA's thinking, but if it succeeded, it would open the door to a far less expensive way to resume the exploration of Mars. The stakes—for NASA, for JPL, for the Pathfinder team, and for the future of solar system exploration—couldn't possibly have been higher.

The failure of Mars Observer led NASA once again to change its plans for Mars. Instead of a series of small landers that would serve as scientific stations around the planet, NASA decided to establish a continuing program of Martian exploration. Pathfinder would still serve as the demonstration project, but it would be followed with a series of satellites to Mars at every launch opportunity—in 1998, 2001, and again in 2003. NASA even began to envision a mission that would bring samples of Mars back to Earth, and a mission that would carry astronauts to Mars.

When Pathfinder became a science mission, as well as an engineering mission, the choice of a landing site became more complicated. Pathfinder would not only have to land safely—it would also have to land in a place suitable for studying geology. While Rob Manning was busy developing the entry-descent-and-landing system, Matthew Golombek, Pathfinder's chief scientist, began work on the choice of a landing site. It was a critical task that NASA hadn't had to face since Viking. Manning's job was to determine *how* Pathfinder would land; Golombek's was to decide *where* it would land.

The basic outline of the problem seemed daunting. The surface of Mars is littered with rocks, debris, sand dunes, craters, huge volcanoes, and yawning canyons. No matter what region of Mars Golombek settled on, he would have to come to grips with terrain that wasn't designed with the landing of spacecraft in mind. Even if he could have found the Martian equivalent of, say, the smooth dry lake bed at Edwards Air Force Base where the space shuttle lands, that would not have solved his problem. Golombek is a geologist, and one of Pathfinder's principal aims was to study geology. He wanted a site with rocks, preferably one with a grab bag of different rocks. If Golombek directed Pathfinder to a site with too many rocks, the spacecraft might become trapped or damaged in the rubble, and he would get no data. If he chose a site with too few rocks, Pathfinder's rover might not be able to get to many.

The search for a landing site was going to be every bit as difficult as it had been for the Viking spacecraft. The Viking team made last-minute changes when it arrived in orbit around Mars, after surveillance cameras showed that the sites that had initially been chosen were probably too risky. Pathfinder's planners would not have that luxury. Pathfinder was not designed to orbit Mars. It would plunge directly into the Martian atmosphere and begin its descent, a direct shot from its launching on Earth. There would be opportunity for course corrections en route, but no opportunity to look before landing.

To make matters even more complicated, there were no images of the Martian surface detailed enough to be able to see how rocky it was. Initially, Golombek had planned to use new photos from Mars Observer. "We had basically no new information beyond what we had from Viking," Golombek recalled after the mission. "What we had was 20 years of people thinking about that data and trying to make sense of it. We were trying to select a site whose characteristics on the order of a meter (about a yard) were important." A boulder that size might, for example, keep the rover from getting off the lander. But the best available photographs of the Martian surface didn't show anything smaller than a hundred yards across. "You're looking at the highest resolution images, and something the size of a football field is the smallest thing you can see," Golombek said. "Mars Observer would have helped tremendously. We would have had information that spanned that scale." When Mars Observer was lost, Golombek and his team were left to do the best they could with the information at hand. They spent two-and-a-half years "looking at all the information we could get, in unbelievable detail, to try to infer what we could."

The first consideration in choice of a landing site was safety—"job number one," as Golombek called it. "You're flinging this expensive piece of hardware at Mars, and if you don't land safely you don't get any science." Golombek had two sets of considerations to keep in mind as he began to search for a suitable landing site—the mission's scientific goals and its engineering requirements. Some of the engineering requirements arose from Manning's entry-descent-and-landing system. The spacecraft, for example, needed at least 55 seconds after the parachute was deployed to stabilize itself and to get rid of its heat shield and protective backshell. That meant that the site had to be at the Martian equivalent of sea level. On a site too far above sea level, the spacecraft wouldn't have time to prepare for landing.

The site also had to be at a latitude at which the spacecraft would be in direct sunlight for most of the day, so its solar cells would get enough energy to recharge the battery. At too high a latitude, two things would happen: The spacecraft would have a bad angle toward the Sun, and the colder temperatures would mean the spacecraft would need more energy to keep warm. That meant Pathfinder's landing site would have to be between the latitudes of 10° and 20°N.

Pathfinder would bounce radar signals off of the ground to determine when to fire its retro-rockets to slow its descent. The signals had to be cleanly reflected for the system to work. "There are certain regions on Mars where radar hits the surface and doesn't come out," said Golombek. "You're going to have this thing trying to measure the distance, and it's not going to

At the Jet Propulsion Lab in the sandbox used
to test the operational scenarios for the lander
and rover, lander camera image mosaics (right &
lower) helped scientists evaluate the configuration
of the rover mounted on the lander petal and gave
them a first color look in stereo at the terrain and
the area directly adjacent to the rover petal. The
lower left image shows details of the surface in
front of the pedal where the forward rover ramp
would be deployed. Images such as these were
used to decide if it was safe to deploy the rover
ramps (ramps were spring-loaded devices that
open abruptly; if they hit a rock, they could spring
back to damage the rover). Operational scenarios
were designed to get the rover off the petal as
soon as possible (lower right), so that the solar
cells beneath the rover could begin generating
electricity.

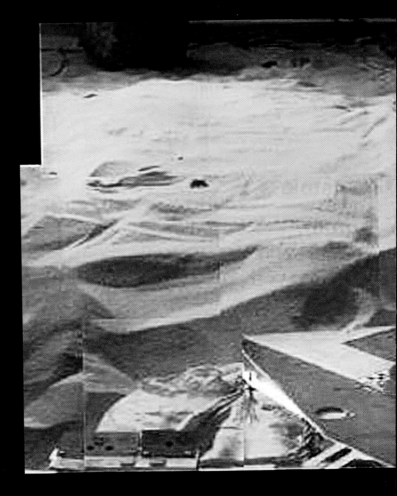

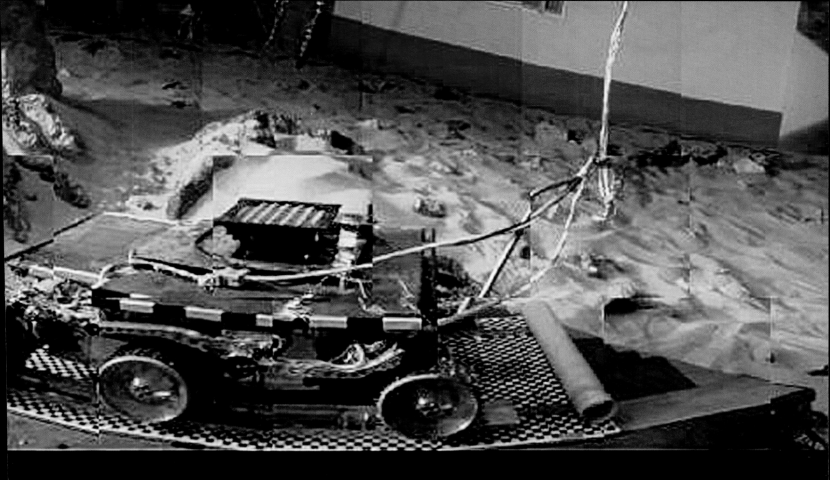

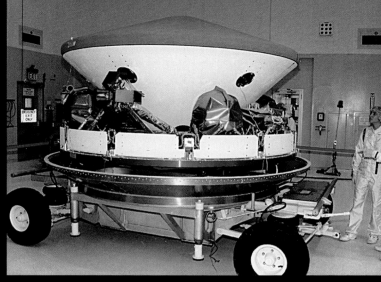

The Mars Pathfinder spacecraft arrives in a cylindrical container in August 1997 at Kennedy Space Center, Florida, after being trucked across the country from the Jet Propulsion Laboratory in Pasadena, California (above, far left). The spacecraft was disassembled, tested, and refitted together. Note the size of the craft, using the people standing next to it for scale, as it rests in a cart with its heat shield oriented downward and the cruise stage and gold-covered propulsion tanks upward. Because of requirements to keep the spacecraft clean to avoid contaminating Mars with

see anything." The cause is thought to be dust cover a few yards thick. If Pathfinder tried to land in such a region, it wouldn't be able to use its radar to determine its altitude, and its rockets could misfire. If the spacecraft drifted over a mesa as it was descending, the sudden change in altitude could produce confusing signals that would cause the rockets to fire at the wrong time or not fire at all—or the spacecraft could smack into a cliff.

"Those are the engineering considerations—you look at all of those," Golombek said. "Then, scientifically, we wanted to select a place where we could do the best science. If you're Viking and looking for life, you might go to a very different place than you would with Pathfinder." Pathfinder's scientific goal was to conduct a geologic analysis of the Martian surface. "What you come down to is—it's a rock mission. You want to look at rocks up close, try to understand their mineralogy and the kind of materials they're made

of. Then you say, 'Where can I go on the planet to learn the most about Mars?'" Golombek decided to look for a site that would have a variety of different rock types. "We wanted a grab bag, a smorgasbord, a potpourri of rocks. The more kinds of rocks we could find, the more we could learn about what makes up the Martian crust."

For geologists, rocks are tools for understanding the past. Pathfinder's rover was designed to enable Golombek and his team to take a very close look at rocks on the surface of Mars—something no previous mission had done. Viking, though it provided wonderful images of the Martian surface, had no capability to analyze what it saw. "We knew almost nothing about what Mars was made of," Golombek said. The only clue at all was a special and unusual collection of 12 rocks known as the SNC ("snick") meteorites. These meteorites are pieces of Martian rock that somehow made their way to Earth after they

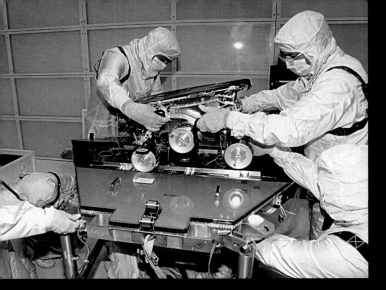

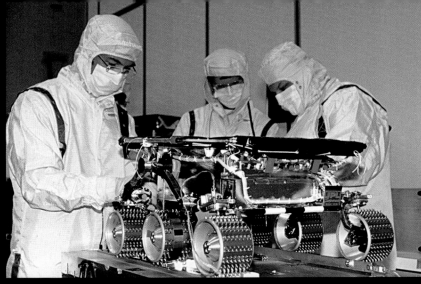

Earth organisms, final processing took place in a clean room (hence, the white suits). During final preparations before launch in late 1997, engineers attach the rover to a lander petal in a compressed or stowed configuration; it must stand up to be in the configuration shown above at right. The metal wheels have cleats for traction. The six-wheel-drive rover has a solar panel on top and a rocker-bogie suspension system, which gives it tremendous mobility. Shown flexing (above, second from right), the rover can drive over obstacles as high as 1-1/2 times the diameter of its wheels.

were blasted from the surface of Mars during meteorite impacts. The rocks have been invaluable in the study of Mars. Until Pathfinder, they provided all the evidence scientists had about the geological makeup of the planet. Unfortunately, however, they are all the same kind of rock—called basalt. "Basalt is relatively low in silica and relatively high in iron and magnesium. It is probably one of the most common rock types in the solar system. On the Earth, all of the ocean basins are made up of it," Golombek said. It is made when lava wells up from a planet's mantle, rises through cracks in the crust, and solidifies at or near the surface. That happens in Earth's oceans as crustal plates underneath them gradually pull apart. The molten rock hits the water and immediately cools or "freezes," becoming basalt. All of the dark areas on the moon are made of basalt. The lava flows and black sands of Hawaii are made of basalt. And all 12 of the pieces of Mars available to scientists are made

of basalt. "So that's all we knew about Mars," Golombek said.

To find a grab bag of rocks, Golombek decided to look for a landing site near the mouth of an outflow channel—an area where one of the ancient catastrophic floods on Mars would have spilled out into a low-lying area, leaving behind an assortment of rocks washed down from the highlands. Such an assortment was likely to include some of the oldest rocks on Mars, dating from about four billion years ago, and the rocks would be the tools for understanding what Mars was like then—a particularly interesting time period, because it is roughly when life began on Earth. If researchers could learn something about what Mars was like at the time, they might learn whether life could have arisen on Mars at the same time it was appearing on Earth. "If you could find a rock that shows evidence of liquid water, you would have a huge way for defining the

PATHFINDER

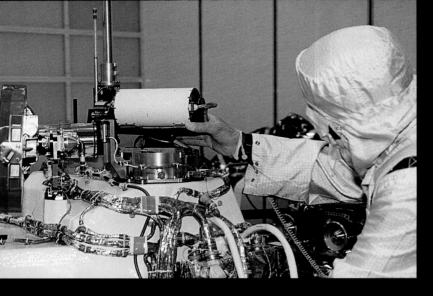

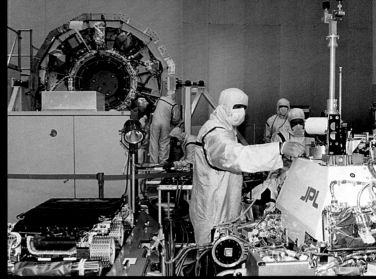

The Pathfinder imager (white canister, upper left) is a stereoscopic camera with spectral filters in the visible wavelengths. The imager worked in its stowed configuration (pictured) as well as on top of a mast about 2.5 feet high (0.8 m) that it deployed after sol 2, which also provided vertical stereo. The rover is shown (upper right) in its stowed configuration on a lander petal. The metallic cylinders are the low-gain antenna and the UHF antenna for communication with the rover. In the lower picture, the spacecraft is attached to the third stage of the launch vehicle. After enclosing the spacecraft in the shroud, it will be attached to the top of the stacked first and second stages on the launch pad. Launching a spacecraft to Mars in a minimum energy configuration can be done only once every 26 months because of the dissimilar solar rotation rates of the Earth and Mars. If the schedule slips during assembly or processing, and the spacecraft is not mated to the launch vehicle and ready to go during the roughly one-month launch opportunity, the spacecraft cannot be launched for another 26 months.

kind of environment that was likely on Mars at the time when life on Earth got started," Golombek said. Although Pathfinder, in contrast to Viking, was not intended to search for life on Mars, its geology objectives might provide important information on whether conditions had ever been conducive to the formation of life. Interest in rocks from the oldest terrain on Mars exploded in August 1996 just a few months before Pathfinder's launch. NASA held a news conference to announce a startling finding: evidence of life had been found in one of the 12 SNC meteorites, a four-pound piece of rock designated ALH84001 that had been found in Antarctica. Almost as soon as the findings were announced, however, an intense scientific debate erupted over whether the strange markings found in the meteorite were indeed fossils or simply the product of some chemical process. If Pathfinder could uncover more information about conditions on Mars at the time that ALH84001 was formed, it might go a long way toward helping scientists resolve that debate.

Golombek began to focus his search on a basin called Chryse Planitia. Chryse (pronounced "crissie") is about two kilometers below the Martian equivalent of sea level, and it adjoins highlands about two kilometers above sea level. Chryse had been chosen as the landing site for Viking 1, partly for some of the same reasons that made it attractive to Golombek—it was low in elevation and not too far from the equator. As he began to study Chryse more closely, Golombek began to look for possible landing sites. In choosing a site, he had to consider what engineers refer to as the landing ellipse. The engineers planning Pathfinder's flight could not say exactly where it would land—they could only say that it would land somewhere inside the landing ellipse—an oval about 120 miles long and 60 miles wide. So Golombek wasn't looking for a safe pinpoint landing site to which Pathfinder could be directed—he needed to find a safe ellipse that seemed free from obstacles. "We looked at all the places in Chryse where you could possibly get that ellipse," Golombek said. "We asked

whether they were dangerous places or not."

Then Golombek looked at temperature measurements from the Viking orbiters. The orbiters had recorded how fast temperatures changed at dawn, as the sun rose and began to warm the planet, and at dusk, as the planet's surface began to cool. Those measurements provided another hint of the nature of the surface. Regions covered with dust, for example, would cool relatively quickly after dark. Areas made of solid rock would hold the Sun's heat longer at the end of the day, cooling more slowly after dusk. Color evaluations of the surface hinted at how much dust was present. Martian dust is red; the underlying rocks are darker and bluer. Redder surfaces were more likely to be covered in dust; bluer surfaces were likely to be rockier. "We didn't want a lot of dust, because it could inhibit the rover, and it could coat the solar cells—and you'd lose power," Golombek said. As a consequence, Pathfinder went to one of the rockiest places on Mars. It was near the mouth of an outflow channel, just as he had wanted—and it was likely to be covered with that grab bag of rocks he was eager to study. "It was estimated that roughly 20 percent of the surface would be covered by rocks, which is pretty similar to what we found at Viking 2," Golombek said.

Some of the many expert reviewers looking over Golombek's shoulder were skittish about that choice. They didn't know whether Pathfinder would survive its landing in such rugged terrain. "For something like the choice of a landing site, everybody in the world wants to get in your knickers and see what you are doing," Golombek said. "There were a lot of reviews, and all of these reviewers wanted to pick it apart. A lot of them wanted to go to the safest place on Mars."

Golombek was confident, however, that Pathfinder would do fine in a rocky area, and he had two reasons for thinking so. The first was that Pathfinder was tested on Earth in conditions very similar to what it would experience during its landing on Mars. Manning had dropped Pathfinder on a surface in the NASA vacuum chamber that was

After a smooth countdown the Mars Pathfinder spacecraft is launched into the night sky at 1:58 a.m. on December 4, 1996, at Cape Canaveral Air Station, Florida. It was the third day of the one-month launch period, with a launch window of only one minute each day. Bad weather had forced cancellation of the first attempt; a problem with non-synchronous computers at the launch site had resulted in scrubbing the second attempt.

much rougher than the Martian surface, and Pathfinder's air bags had shown they could take the punishment. "We ended up making the most robust lander that's ever been designed to land on a planet," Golombek said. "Pathfinder could land anywhere."

Additionally, Golombek knew what to expect on Mars, because he had studied and walked across an outflow channel on Earth that was remarkably similar to the one he had chosen on Mars. It's in an area called the Channeled Scabland in central Washington State. About 13,000 years ago, during the ice age, part of an ice sheet that then covered the region formed a dam, creating a huge lake in western Montana. The lake contained as much water as Lake Erie and Lake Ontario combined. Toward the end of the ice age, the dam suddenly broke, and that immense quantity of water roared downhill to the Pacific Ocean, completely emptying the Montana lake in about two weeks. "It was the kind of thing man has never seen," Golombek said. It carved streamlined islands and channels across Washington State, leaving a trail of debris behind it—exactly the kind of thing that happened at Golombek's proposed landing site. "This is small by comparison to Mars, but you can look at it and learn a lot about how this process works." Parts of the Channeled Scabland, where water had collected, are almost perfectly flat. A large airport was built near Moses Lake in the Scabland, and the surface was so flat that builders did not have to grade the land—they simply laid the runway on the existing surface. Near the end of what's called a "depositional fan," Golombek found "rocks as big as your house"—not an ideal landing site even for a spacecraft as sturdy as Pathfinder. "We used this geologic analogue as a way to infer what the surface would be like on Mars," Golombek said. On one field trip, he invited teachers and planetary scientists to accompany him, and he brought along Pathfinder's rover for a test drive.

The depositional fan he studied in the Scabland is called the Ephrata Fan. Using his research there, Golombek was able to put together a pretty good picture of what the rock distribution would be like in the Chryse basin. He wanted to land near an outflow channel in Chryse called Ares Vallis, which resembled Washington's Ephrata Fan, except that Ares Vallis was much, much larger. "It is what you would get if all of the water in the Great Lakes cut a channel to the Gulf of Mexico in about two weeks," Golombek said.

Finally, Golombek determined that Ares Vallis would indeed be the landing site. In making that choice, he was up against some discouraging history. Viking 1 was originally supposed to land on Ares Vallis, but engineers rejected it as a suitable landing site after reviewing images taken by Viking 1 from orbit. "They said it was too rocky, we can't deal with it, we've got to go somewhere else. So when I first proposed this, the skepticism was enormous. The reviewers—some of whom were veterans of Viking—said, 'We rejected this place, and you're going to go back there and kill the spacecraft—and kill your career.'" Golombek held fast to his choice, enduring "more meetings and reviews that you could believe."

In the end, Golombek and his team made three predictions about what would happen when Pathfinder arrived at Ares Vallis. The first was that the spacecraft would land safely, that the area wouldn't be too rocky and the radar would accurately sense the planet's surface. The second was that the surface, although rocky, would be smooth enough for the rover to make its way over the terrain. And the third was that the area would be less dusty than either of the Viking sites.

"All three came true," Golombek said. "We have now evaluated all the data we used to infer the properties of the site, and everything we used did a great job." Viking's planners were wise not to land at

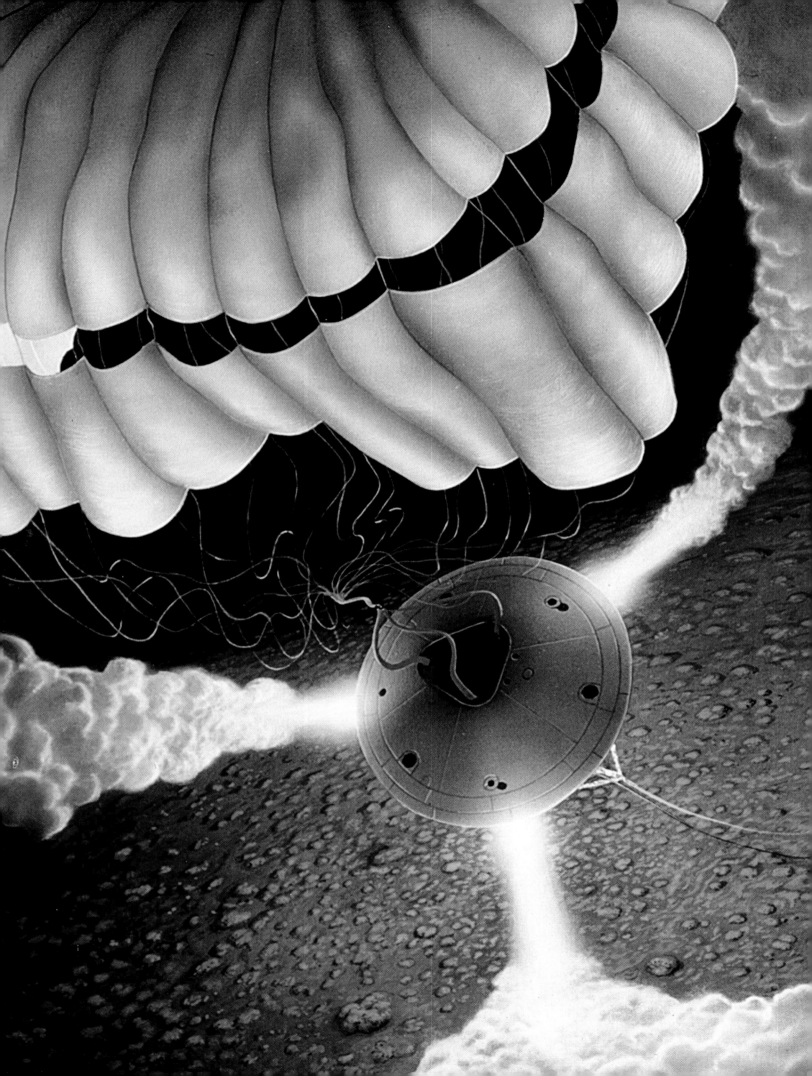

Painting illustrates Pathfinder's landing on Mars. The air bags surrounding the lander inflated, solid rockets on the inside of the backshell fired, and the lander's tether was cut. Pathfinder bounced more than 15 times on one of the rockiest places on Mars without a single puncture.

Ares Vallis, Golombek said. Viking needed a clear area to put down its three legs, and Ares Vallis did not offer too many such spots. "They were fortunate," Golombek said of his Viking predecessors. "They did the right thing by waving off the Ares Vallis site. There's no way I would have wanted to land Viking there. But with our air bags, we could land there with impunity."

Once the Pathfinder site was chosen, Golombek could begin looking forward to the geological reports that Pathfinder would send back from Mars. Even before the scientific reports began, however, he felt that an important milestone had already been achieved. "The landing site choice consumed me for two-and-a-half years. That was, without a doubt, my greatest satisfaction. We just nailed that site."

Richard Cook started work on Pathfinder in 1992 as the mission designer. Among the first things he had to do was to decide on the spacecraft's trajectory—the path it would follow to Mars. "A lot of the spacecraft design begins to come out of that, the geometry, how far you are from the Earth and the sun—it affects your power, all kinds of things," he said later. "Initially we had a design that took a much longer time to get there, and there was real pressure to shorten the flight time." In addition, because Pathfinder was intended to be an engineering demonstration project, the mission team wanted to be able to get data back from the spacecraft continuously. "We wanted to be sure that if it didn't work, we could understand why it didn't work....In order to do that, you had to worry whether you could see the spacecraft when it was going through the atmosphere and whether you could build an antenna system on the spacecraft that allowed it to communicate to you all the time." The original trajectory made that difficult; Cook had to change it. With the new design,

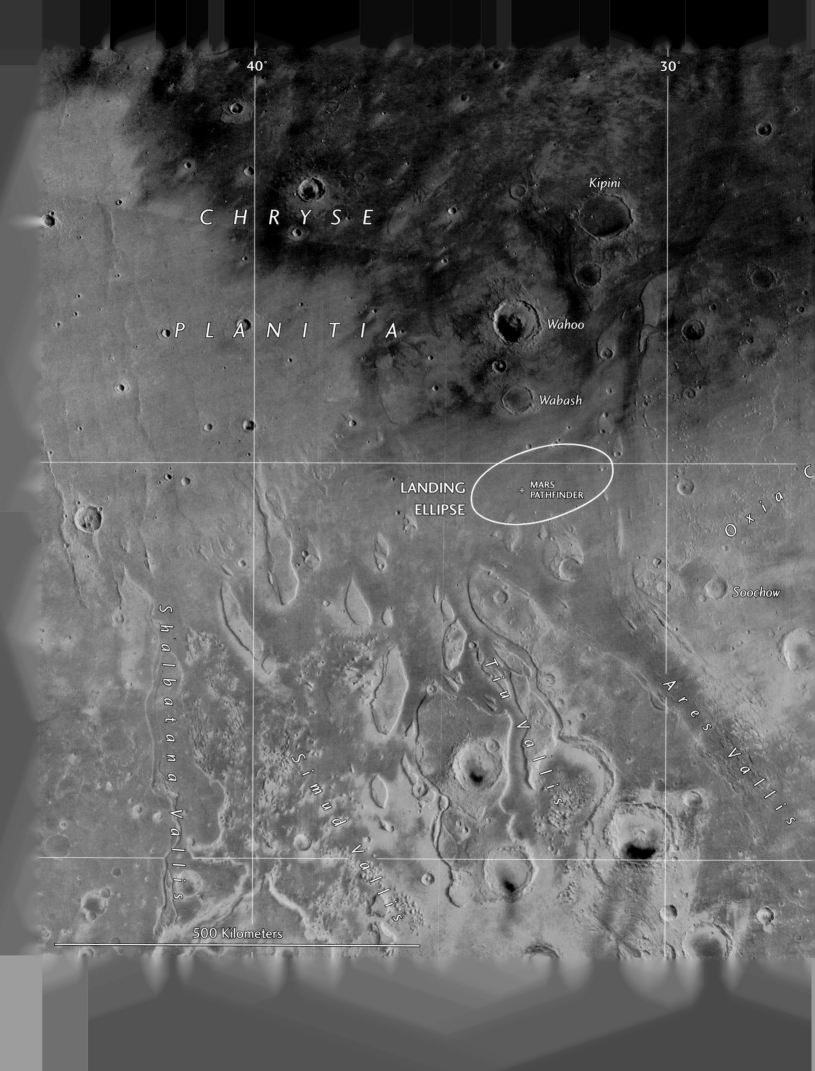

C H R Y S E

P L A N I T I A

Kipini

Wahoo

Wabash

LANDING
ELLIPSE
+ MARS
PATHFINDER

O x i a

Soochow

Shalbatana Vallis

Simud Vallis

Tiu Vallis

Ares Vallis

40°

30°

500 Kilometers

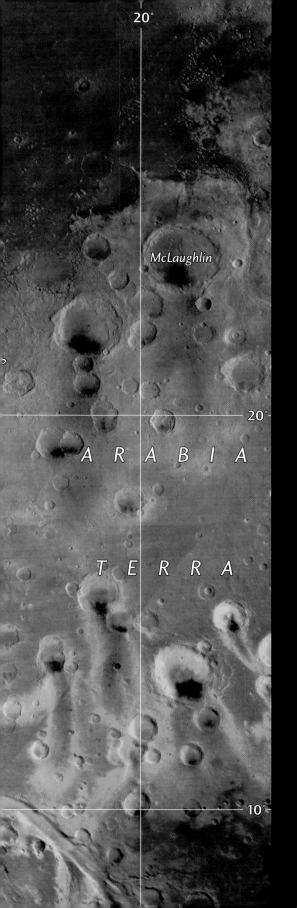

20°

McLaughlin

20°

A R A B I A

T E R R A

10°

Color regional mosaic (left) shows the 120-by-60 mile (193-by-97 km) landing ellipse targeted for Pathfinder. Various uncertainties in the location of the spacecraft and its navigation and the state of Mars's atmosphere precluded a more precise target. The ellipse is downstream from the mouth of a giant catastrophic outflow channel called Ares Vallis, which formed when a flood carrying roughly the volume of water that now fills the Great Lakes deposited material from the heavily cratered terrain to the southwest. Certifying that the site was safe for the Pathfinder landing system proved difficult because the highest-resolution images available showed features only the size of a football field. To determine where the lander came to rest, scientists compared transmitted images of features on the horizon with features in Viking orbiter images such as this (below) taken more than 20 years before.

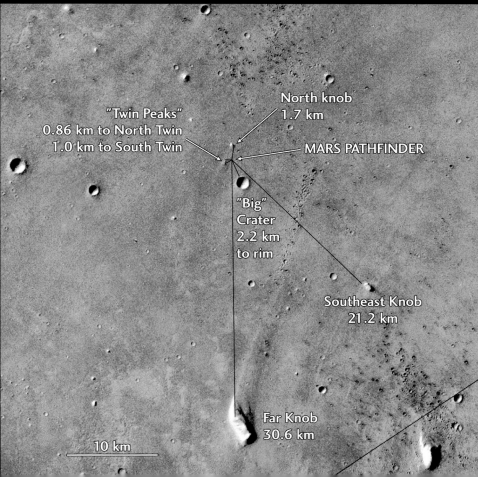

North knob
1.7 km

"Twin Peaks"
0.86 km to North Twin
1.0 km to South Twin

MARS PATHFINDER

"Big"
Crater
2.2 km
to rim

Southeast Knob
21.2 km

Far Knob
30.6 km

10 km

On the surface of Mars at Ares Vallis on sol 1, July 4, 1997, the Pathfinder imager made this first color mosaic of the stowed rover, dark rocks, red dust, and pale sky. Air bags inhibited deployment of the rover ramps (the gold, silver-edged cylinders at either end of the rover). The petal had to be raised and the air bags further retracted before

plates with color targets mounted on the spacecraft at left are magnetic targets. Dust free in early
images such as this, the four strongest magnets
had attracted dust of the same color as in the
atmosphere after 70 sols, which proved that airborne dust on Mars is highly magnetic. Twin Peaks
in the background, just over half a mile (one km)
to the west-southwest, contributed to the rapid
location of the lander in Viking orbiter images
(see pages 140–141).

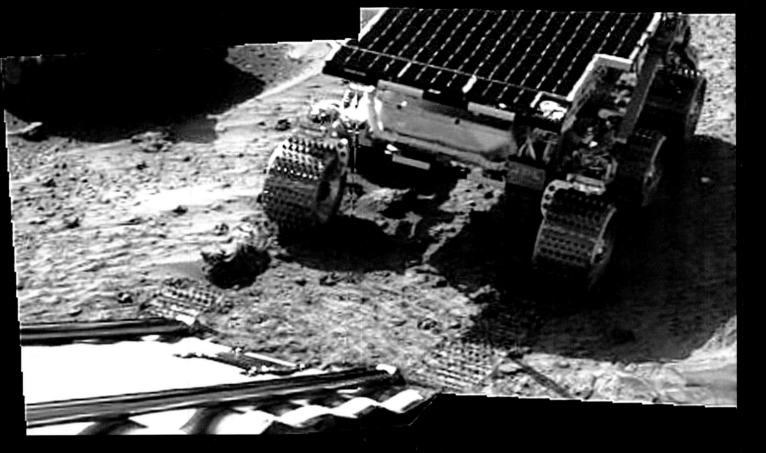

Image from the lander camera on sol 2 shows the rover successfully off the rear ramp and onto the Martian surface. The rover wheel cleats leave clear indentations, indicating a very fine-grained and compressible soil. The rover placed the chemical analysis instrument against the soil for the first in situ measurement of surface materials at the end of sol 2. Jubilant scientists and engineers in the Pathfinder mission control room at the Jet Propulsion Laboratory in Pasadena, California, celebrate the successful landing and surface operations. Pictured are Rob Manning, (below, right), flight system chief engineer; Leslie Livesay, (below left), telecommunications system manager; and, standing next to her, Matt Golombek, Pathfinder project scientist.

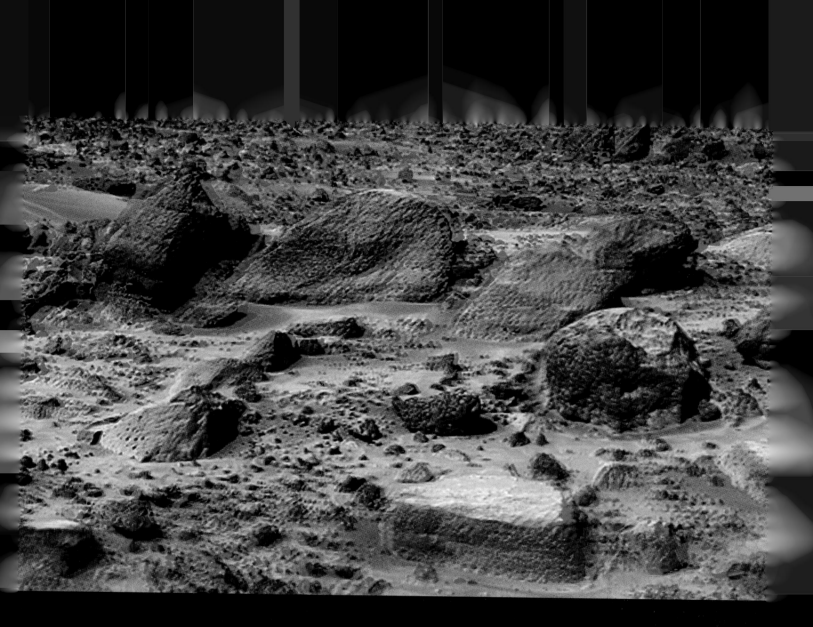

When viewed with the red/blue anaglyph glasses (found in the pocket of the inside back cover) with the blue lens to the right, the terrain as seen from the lander becomes fully three-dimensional. The stacked or layered rocks (in the Rock Garden, above) provide evidence for deposition in high-velocity flow, as would be expected in catastrophic flooding thought to have occurred some two billion years ago.

One half mile away from the lander, the Twin Peaks (pages 146 and 151) stand some 150 feet tall and provide a frame of reference for scientists at the Jet Propulsion Laboratory trying to determine exactly where Pathfinder landed on July 4, 1997.

One day later, the six-wheeled rover named Sojourner took its first spin on the red planet (pages 147-150). The Pathfinder mission, which marked man's return to the surface of Mars after an absence of more than 20 years, heralded a new era of exploration characterized by less expensive projects that could be undertaken more frequently. There are seven more missions to Mars planned for the next several years.

Sojourner passes Barnacle Bill (page 152) on its way to explore the dusty, boulder-strewn Martian landscape. The rover would circumnavigate the lander for 83 sols before communications were lost at the end of September 1997.

Pathfinder's arrival date at Mars was July 6. When Cook and his colleagues realized how close they were to Independence Day, they made a slight change: They rescheduled the landing for July 4.

"Viking was originally going to land on July 4, and that was part of why we wanted to do it. Not to tweak Viking. But there was always a bit of a feeling, Damn, I wish we could have landed on that date," Cook said. Viking 1's landing had been pushed back to July 20, 1976, missing the nation's Bicentennial celebration, when mission planners decided at the last minute to change its landing site. To keep the mission as simple as possible, Pathfinder, unlike Viking, would not orbit Mars before landing, so there would be no opportunity to change the landing date after launch. "Because we were going straight in, we knew we were going to land on July 4—one way or another," Cook said.

NASA engineers are accustomed to exhaustively exploring every possible option before making design decisions, but that was not to be the case for Pathfinder. Because of the limitations on time and money, Pathfinder's designers had to make decisions quickly. "We didn't have enough time to really take all of the different paths to some logical conclusion," Cook said. "We had to make decisions with a relatively small amount of information." When problems developed later, engineers had to fix them. They didn't have the option to back out and start again with a new design. "Viking had the advantage that they got delayed by two years," Cook said. "I'm sure they wouldn't call it an advantage, but they spent two years looking not just at the design options they chose but also at a lot of different ones. It was a much more reasonable and rational process. But in this era, you need a certain amount of pig-headedness: You say 'this is the way we're going to go,' and you make that work.

"That happened over and over again. Often we thought that maybe something wouldn't be the right option if we went back and tried to optimize the whole thing, but we had to make it work. It gave everybody real motivation." As the team bulled through problems, its confidence rose. "Things began

to go more easily, and nobody really got disheartened by it. That was a hallmark of the team." Not everyone was convinced, however. NASA headquarters in Washington and the experts reviewing the project often had doubts. "We had our big review in July of 1993, and there was a fairly large subset of the review board that thought that this mission just couldn't work," Cook said. That was a year-and-a-half into the project, and it was a harrowing experience.

The Pathfinder team was a much more tightly integrated group than was typical of earlier missions, Cook said. Jobs were shared, communication was encouraged between, say, spacecraft designers and the people who would actually operate the mission. That gave the mission operators the chance to suggest changes that would reduce the cost of running the mission. Individuals were often given more responsibility than they might have been given in other missions. "We had a lot of problems with communications, because the spacecraft is sitting there on Mars and everything is moving and the receiver on the lander is going on and off to save power. When problems like that happened, we'd have to come up with a response in real time, build the set of commands and sequences that implemented that, go test them in the test bed, and then uplink the spacecraft and wait for something to happen." Cook would make the decision with the missions operator on duty, and the operator would send the commands himself. "I'm pretty sure that in the history of JPL that has never happened—that one person has done all of that," Cook said.

Pathfinder was launched from Cape Canaveral aboard a Delta II rocket at 1:58 a.m. on December 4, 1996, on its seven-month journey. Immediately following the launch, Cook was forced to grapple with the problem of the spacecraft's malfunctioning sun sensor and with a frustrating and almost fatal communications problem. Once the spacecraft was on its trajectory to Mars, the mission proceeded smoothly. The Pathfinder team then had time to think about the next critical phase of the mission: the entry-descent-and-landing sequence that Rob

Manning and his colleagues had worked on for so long. To reassure himself that the system would work as planned, Manning had commissioned an unusual final test just before launch. Pathfinder, like many of its predecessors, relied heavily on what the engineers call "pyrotechnic devices." That's a technical term for fancy firecrackers timed to explode at key points in the sequence. These devices are used to blow bolts when parts of the spacecraft need to be separated, or to open hatches, release parachutes, and perform other similar chores. Years of experience have shown that these electronically triggered firecrackers perform exceedingly well. Problems can occur, however, when they fire at the wrong time—which can be the result of a computer error.

The entry-descent-and-landing sequence would last only four-and-a-half minutes, and during that time 41 pyrotechnic devices would have to be triggered in precise sequence to assure that the spacecraft made it safely to the ground. Manning commissioned a test in which each of those firecrackers was disconnected from its cable, and the cables were connected to an external computer which could monitor the timing of the triggering sequence. The tangle of wires used to connect the external computer was called the octopus. "This cable was lying all over Pathfinder inside and out. It was just a mess of wires," Manning said. (The octopus is now in a box underneath his desk—an "oversize memento," as he calls it.) Manning ran the test over and over again just before launch, hoping to uncover any computer glitch on the ground, not while the spacecraft was hurtling toward the rocky Martian surface. "I wanted to watch each one, on the actual spacecraft, running the real software," Manning said. There was some risk: if the cables were not disconnected before the test, Manning risked blowing a device that could not be repaired in time for launch. And if the cables weren't properly re-connected after the test, the spacecraft would fail during landing.

Manning thought the test was crucial. "Our team's mission was to come up with a process where none of us had to be worried on the third of July. We wanted

to get a system that was as reliable as Viking—or more so, if possible. That was our goal."

When Pathfinder was launched, however, the software testing had not been completed. "The way software works is—you're never done," said Cook. "You can always test software more; there are always things that go wrong." Three or four months before launch, Cook realized that the software was not going to be ready in time. So his team came up with a contingency plan: They would race to fix as many of the software bugs as they could before launch, concentrating on the software for the "cruise phase" of the mission—the travel between Earth and Mars. Problems with software needed for landing and for spacecraft operations on the surface would be fixed and sent up to the spacecraft while it was on its way.

Software testing continued at JPL. When changes were made, they were rigorously tested on the ground. Adjacent to the control room at JPL was a large room that had been converted into a sandbox. Inside were a mockup of the lander and the rover, where changes in software commands could be tested. "We found a lot of problems, so it was good we did it. And we created fixes that we were going to uplink to the spacecraft before it landed," Cook said. In mid-June, about three weeks before landing, the fixes were radioed to the spacecraft.

The four-and-a-half-minute landing sequence, with all of Manning's pyrotechnic devices exploding in turn, was to be handled entirely by the spacecraft. A signal to begin the sequence would be sent from Earth, and the spacecraft would be on its own until it landed. During the final software tests, engineers decided that they should close the valves on various tanks and thrusters to prevent any leaks. The initial plan had been to leave these so-called latch valves open, but upon reflection it seemed wiser to close them, Cook said. That change was sent along with others about three weeks before landing.

At 8:30 in the morning on July 4, Pathfinder was a little more than an hour-and-a-half away from landing on Mars. Tension was mounting in the control room as engineers reached the point at which they would no longer be able to communicate with Pathfinder until after landing. Then a problem appeared. "We're watching data come back from the spacecraft, and the latch valves don't close. We had run a test a month beforehand where the same thing had happened, and it turned out we had been running the wrong version of the software. We thought, 'We can't let this happen on the spacecraft—we've got to make sure we get the right version of the software on the spacecraft.'" Leaky valves could allow gases or liquids to escape, depleting fuel or throwing the spacecraft off course.

It now seemed that Pathfinder, in the critical moments just before landing, might be running the wrong version of the software—possibly a version with errors that hadn't yet been corrected. There were only minutes left in which any changes could be made. "We were in the situation for real, and the basic conclusion we came to is that we were running the wrong version of the software. We had made all these changes to make it work—but now somehow we had the old version on the spacecraft. We were right at the point where we wouldn't be able to send commands." At that point, Pathfinder actually held two copies of its software. Cook needed to make an immediate decision: Should Pathfinder switch to the other version of the software on board? Or should it stay with what it had? "My answer to that was no— we're not going to do anything. We're going to ride this out." But Cook and the team immediately went to work to see if they could determine whether Pathfinder was running the correct version.

But how could they tell? They had to find some change that was in the correct version of the software and not in the previous versions, and see whether that change was in the software on the spacecraft. They started poring over all the changes that had been made. But they found nothing they could use to quickly identify the software on Pathfinder. Then they began focusing on something Pathfinder's engineers had done as a joke. Pathfinder was reporting regularly to the ground on mission status, and when things hadn't changed from the last report, the mes-

sage from the spacecraft would be mostly composed of what are called "fill bits"—meaningless information that was thrown away on Earth. Pathfinder's software engineers had put their names in there, so the reports would transmit their names over and over again during the flight. Just after launch, a JPL mechanical engineer named Jordan Kaplan had died in a plane crash, and, as a tribute, the software designers had incorporated his name into the fill bits. Kaplan's name was in the current version of the software, but not the older versions. That was the tipoff Cook had been looking for. If Kaplan's name was in the messages coming from Pathfinder, then it was indeed running the right software.

"We saw that list of names come down every minute," Cook said. "We usually threw it away, because we didn't care. But it was on the ground, and we could see it. So we went to Glenn Reeves, the lead flight-software engineer, and Jesse Wright, the lead ground-system engineer. We asked if they could show us this fill data. Reeves started fiddling around, because none of us knew how to do this—he was the only one. He printed it out on the screen, and sure enough, Kaplan's name was in there. We knew we were running the new version of the software. We ran back into the control room and told them, and everybody was relieved. We dodged a bullet there." Cook's decision not to have Pathfinder switch to the other version of its software had been the right one.

That brought Cook back to the original problem. If Pathfinder had the right software, why hadn't the latch valves closed? After thinking about the problem for a while, Cook and his team realized that the software testing on Earth was not duplicating exactly what was happening on the spacecraft. On Earth, when software changes were made, the existing software was erased and the new software installed. On Pathfinder, the new software was simply written over the old software—the old software wasn't erased first. Engineers realized that not everything was being over-written. In some cases, Pathfinder was refusing to accept some pieces of the new software because it already had them. "It was a breakdown in the way we

On sol 33, the rover imaged the rock called Ender (nubby-textured, foreground) and the lander. Air bags show prominently. Yogi rock bulges beyond the lander. The JPL logo, on the side of the electronics box, is visible adjacent to the American flag. Looking back (opposite), the rover viewed the lander, the rear ramp, and its own tracks.

tested it. We didn't test it the way we were going to fly it," Cook said. The valves remained open, but they didn't do any harm.

Pathfinder's entry-descent-and-landing sequence began, and communication with mission control was lost. For four-and-a-half anxious minutes, mission controllers waited for a signal from Mars indicating that Pathfinder had made it. The Deep Space Network, a series of ground stations around the world, pointed its antennas at Mars, and waited. Manning was the flight director during landing, and it had been his job to make sure everyone would be calm before landing. How did he feel? "I would say I was one of the more relaxed people," he said later. "But 'relaxed' is a relative term. I had had clammy fingers for a year, but I felt confident nobody could propose things differently than we did. We did everything we could do. If you had offered me 10 million dollars to do more tests, I would have said I can't think of any."

Five years after work on Pathfinder had begun, the team, now a tightly-knit group, watched for the signal from Mars. Many had jammed into the control room. A weak signal was detected—but it was dismissed as noise. Several seconds passed, and Manning got word

that the Deep Space Network may have found something. "That's a very good sign everybody," he said, and a moment later he got confirmation: "We're getting a firm signal." The control room erupted in a series of cheers, raised fists, embraces—and tears. The five-year effort had paid off. "We're there!" somebody shouted. For the first time in over 20 years, American was back on Mars. The scene in the control room, quiet and tense a moment ago, now resembled the locker room of a World Series winning team (minus the champagne, which doesn't mix well with control-room electronics). Manning had to struggle to restore order.

"Although we're very happy up here, we can't go home yet," he said. One more confirmation was expected from the spacecraft. "It could very well be that the spacecraft has turned off its transmitter at this time, as it would be expected to after it rolls to a stop," Manning announced, in his official role as flight director. "We are now in a blackout situation where the spacecraft will essentially begin to retract air bags. This process takes a little over an hour. We may see a signal when that finishes up. We will keep our eyes peeled of course." The second signal came in on schedule, prompting another round of cheers, shouting, and embraces. "We've got a mission, we've got a mission," said Golombek, who was still waiting to see what his landing site looked like. Manning's entry-descent-and-landing system had worked. After all the doubts expressed by reviewers, after all the testing, after all the work, the system had performed ably. NASA and JPL had shown that they could land a spacecraft safely on the surface of Mars. In his role as designer of the descent phase, Manning's role was over. Tony Spear leaned over to Manning and, affectionately, told him he wasn't needed any more—he was fired. "This is Rob Manning, and I'm out of a job," Manning announced in the control room. "I'm going home." ("Actually, I didn't go home, but I could have," Man-

ning said later. "I'd been up for a pretty long time.")

The engineering demonstration had been completed, but for the Pathfinder science team the mission was just beginning. The rover team, too, was eagerly anticipating the first test of its creation. But before the surface mission could begin, controllers on the ground had to be sure Pathfinder was in the right position and wasn't damaged. They needed pictures from the surface. Pathfinder's camera—the Imager for Mars Pathfinder, or IMP—was readied for the first pictures. At a press conference at JPL following the successful landing, the designer of the camera, Peter Smith of the University of Arizona, provided a vivid and eloquent description of what it

This image of the sky as viewed from the Pathfinder lander, looking toward the sun before sunrise, reveals early morning clouds over Mars. The suspension of very fine-grained (micron in size) dust in the atmosphere colors the sky reddish. The dust appears to be an iron-rich weathering product. The blue color of the clouds probably comes from the forward scattering of light by the small water-ice crystals that freeze out on the dust particles during the night and sublime away by mid-morning. Wispier white water-ice clouds also proved common in the mornings.

was encountering on the cold Martian surface.

"The eyes of the camera are our eyes, and in that sense we are all on Mars," Smith began. "We are there together. You might say that the people of Earth are the soul of this robot. So for the first presentation of images, forget about the engineering and scientific aspects. They're very important, but open your imagination to the experience and beauty of landing on Mars.

"Pretend that you are in the position of the camera. You've awakened in darkness after a rather bumpy landing. You slowly come to consciousness and remember that you are no longer on the Earth. Your name is IMP and you've been hibernating in space for seven months. The ride wasn't comfortable. Your forehead was pushed up against the solar panels. The rover was crammed against your left ear. The high-gain antenna was mashed against your right ear. And that parachute canister—straight down the spine.

"The scientists and engineers, both United States and German, that built you, promised that you would land in one of the most exotic places in the solar system, the flood delta of an ancient valley named Ares, a canyon so huge that it was capable of carrying 1,000 times the flow of the Amazon River. Dry for billions of years, the delta is now an ideal place to look for the diversity of rocks that contain the history of early Mars. Perhaps you will see some evidence…of the dry landscape of the ancient water channels left from the last flood. For your first look, you are sitting close-legged on your petals, coming out of your seven months' meditation, and you start to raise your head, and this is what you see..." At that point, Smith displayed the first mosaic from Mars, showing the first view of the rock-studded, red plain that would become familiar to millions around the world by the end of the Pathfinder mission. The hundreds of reporters assembled to cover the mission broke into applause.

Within the first few days following Pathfinder's landing, Smith was showing reporters panoramic views of the landing site, and the images were being made available on the World Wide Web. On Monday, July 7, the first time since the landing that most people would have been at work, the Pathfinder web site registered 47 million hits as people sifted through a wide selection of color images of the Martian surface. During the first 30 days of the mission, Pathfinder would log a staggering 566 million hits, making the Pathfinder mission the most widely attended Internet event in history.

When the first images appeared on computer screens in the control room, Golombek was ecstatic. The very first look told him that all the work on the landing site had been worthwhile—what he saw was exactly what he wanted to see. "You want a landing site?" he shouted to his colleagues. "I deliver!" Golombek, not ordinarily a betting man, had made two wagers before the flight. One was with a colleague who, near the beginning of the project, bet that Pathfinder would never get off the ground. "He said, 'There's no way, given that amount of money, that you're going to launch a payload with a rover, an imager, and a weather station.' That wager was for a fine bottle of wine. I got it a week after landing, at a geology meeting in front of 40 other people. It was pretty cool," Golombek said. The other was with a geologist colleague, Ray Arvidson of Washington University in St. Louis, who had worked on the Viking mission. Arvidson bet that the landing site would look just like Viking 1, with rock outcrops scattered here and there. Golombek bet that all the outcrops would be buried by the rocks and sediment washed down by the ancient catastrophic flood. The wager was a glass of beer—and Golombek won that one, too.

When the excitement over the successful landing died down, mission planners discovered they had a problem that needed solving before the mission on the surface could get under way. Those first images, which thrilled scientists and the public alike, showed that Pathfinder's air bags had not fully retracted. Billows of material surrounded the lander "petal" carrying the rover—and it looked like they could block deployment of the rover's landing ramps and prevent it from getting to the Martian surface.

The rover positions its alpha-proton X-ray spectrometer against Yogi rock (above). Measurement of its composition took about ten hours. Notice the blue-red color slanting across the right side of Yogi; colors differ in various solar illumination geometries. In a super-resolution view (opposite) Yogi appears dustier than other rocks analyzed by the rover, which may explain Yogi's slightly lower silicon content measurement (the dust has less silicon than the rocks).

Upon landing, Pathfinder had bounced as high as 50 feet and continued to bounce between 15 and 20 times on the Martian surface before it came to rest. The air bags were then deflated, and Pathfinder's petals opened. As it happened, Pathfinder had landed on its base petal. If it had landed on one side or another, the opening of the petal on that side would have been used to straighten the spacecraft so that it was resting properly on the surface. But the retraction of the air bags was a somewhat uncertain process. There was a risk that the bags could get caught on the spacecraft or somehow interfere with its operation. "It was a relief to see the signal come in at that point, and basically everything was perfect," Cook said. Pathfinder sat squarely on the ground. "The tilt was about 2 degrees."

The first thing Pathfinder had to do was a sun search. That is, the spacecraft had to use a camera to locate the sun. The position of the sun would tell Pathfinder which direction it was oriented in. It needed that information to be able to point its high-gain antenna at Earth. "That was an area where we were very concerned it wouldn't work," Cook said.

"It was autonomous, so that meant it was intrinsically risky, because it didn't have any people in the loop." It had been difficult to test the system on Earth, where the Sun is so much brighter. The flight team sent the spacecraft a command to do its sun search just before leaving to give a press conference. "When we came back, at the right time, the Deep Space Network started looking for the signal from the high-gain antenna, Cook said. And there it was, exactly at the predicted signal strength. It was amazing—it was pointed to within one degree of the Earth." The high-gain antenna could transmit far more information than the low-gain antenna that Pathfinder used first to let engineers know it had landed. Proper operation of the high-gain antenna was essential in getting Pathfinder's images back to Earth.

The rush to get the first images back from Pathfinder went beyond the anticipation of looking at the Martian surface for the first time in almost 21 years. "The reason why we were so interested in getting those images first was to get ready to deploy the rover," Cook said. "We needed to get it off the

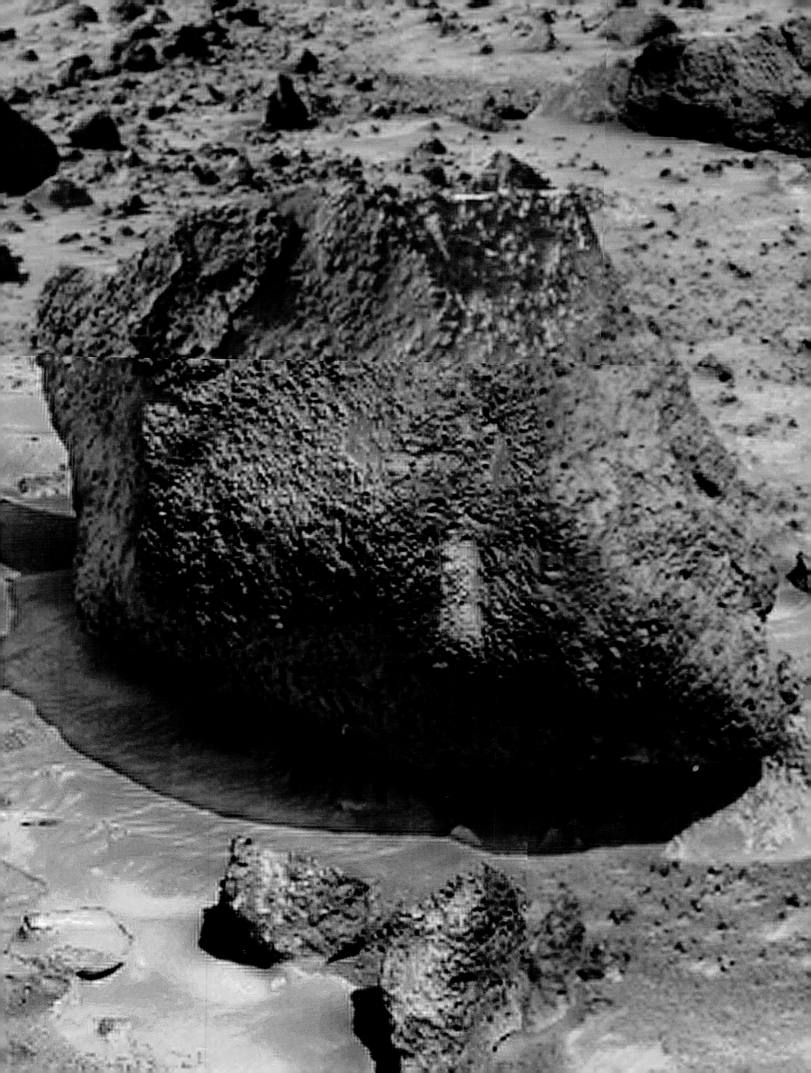

spacecraft because it couldn't do anything until it got off. And it was also sitting on some of the solar arrays. It was blocking the power." The team wanted pictures on either side of the rover to see whether it was safe to unfurl the ramps that the rover would use to get to the surface. There were two ramps—so the rover could leave the lander by going forward or backward. "We didn't want to deploy the ramps until we were sure we didn't have a rock there," Cook said.

When the first images came back of the rover sitting on one of the lander's petals, flight controllers saw the airbags billowing out around the rover. "We were extremely confident that we could deal with it," Cook said, but tests had to be run at mission control first to devise the proper sequence of commands. "Normally, the way we like to run missions is have everything planned out, and have all the commands ready to go, so you don't have to do everything on the fly. But because of the uncertainties, we didn't have everything built up in advance. We were joysticking this mission as much as any mission I've been on in a while." By "joysticking," Cook meant that much of the work of directing the spacecraft was being done in real time, by engineers with their hands on the controls. The petal was lifted 45 degrees, the airbag retractors were operated for 10 minutes, pulling the air bags in, and the petal was lowered. The maneuver worked. The air bags were out of the way.

Then it was time to send the rover on its way. The ramps were unfurled on sol 2—the spacecraft's second day on Mars. "We took a set of images, deployed the ramps and took another set of images. The forward ramp stuck out like a diving board—that wasn't something it was supposed to do." But because the other ramp looked fine, the team was optimistic that it could dispatch the rover that day.

Before the team could relax, however, yet another problem cropped up. The rover and lander communicated by radio modems, not unlike the modems desktop computers used to communicate over telephone lines. But they seemed to be having a problem talking to one another. "We actually got a bunch of information on that first low-gain session that the

rover was O.K.," Cook said. "But as the day progressed, we saw that the rover modem was resending the same information over and over again. It is designed to do that if it detects that it's failing to get data or if its data is not getting through. We began to notice that the number of these attempted retransmits and failures to communicate between the rover and the lander were increasing. So the last two passes of the first day, we got no information from the rover at all. The rover guys began to freak out, wondering whether something was wrong with the modems."

A meeting was called to try to determine the problem. "We weren't ever saying this thing is doomed, but we were confused enough that people were starting to have doubts," Cook said. The problem was serious enough that if it couldn't be corrected it could potentially cripple Pathfinder's scientific mission. "We had had a lot of small failures either in tests or in reviews, so the team had developed a real resiliency to those problems. There was never a time when anybody said, 'We're doomed, this is not going to work.' People were always hopeful we'd figure something out, because we'd figured out a billion things before."

The next morning, when the daily communication with the lander was established, the rover was working fine. Pathfinder's scientists and rover team "were euphoric," Cook recalled. Amid that euphoria, however, mission controllers noticed that the computer on the lander had been reset—the equivalent of rebooting a personal computer. The cause seemed to be some problem with the computer's software, but the ground control team couldn't determine exactly what the problem was. One theory is that the computer became overloaded as it tried to communicate with the rover. "Every time it tried and failed, that would end up swamping the flight computer. The software would say 'I can't keep up, so there must be something wrong,' and it would reset and pretty much do nothing." Just as quickly as the problem appeared, however, it disappeared—the resetting of the computer apparently cleared it up. "To this day, I'm not sure we know what happened," Cook said. "That was a theme of surface operations." Problems would occasionally appear, the cause would be unclear, but somehow the problems would be rectified. "We had several problems where we didn't know what happened," Cook said.

By the end of that second day, the team had driven the rover off the lander, and they were displaying a series of pictures showing it moving down the ramp and leaving its first tracks in Martian soil. For Cook, that was the time he felt he could finally relax. Others had felt that sense of relief at launch or upon landing, but Cook wanted to see that rover on the ground. "That was when I thought, hey, we're there. This is going to work."

The lander was christened the Carl Sagan Memorial Station, in memory of Sagan, who died shortly after Pathfinder's launch. And the rover, which had earlier been named Sojourner (after Sojourner Truth, the former slave who became a distinguished lecturer on abolition and women's rights) began its explorations. Pathfinder scientists dropped any pretense of using coordinates or any other scientific designation to identify the rocks they wanted Sojourner to explore. Instead, they adopted a practice Viking scientists had used, giving rocks and other features whimsical names. Communications continued to be a tricky problem. A week after landing, ground controllers mistakenly tried to communicate with the lander before it turned on—or "woke up," as the engineers say. Hearing no response, they thought the lander wasn't getting their signals, and they didn't make another attempt. Later they discovered that the lander had in fact turned on and was awaiting a signal. An entire day was lost. "We made a little mistake," Cook said at a news conference afterwards. "As the team joked, on the seventh day we rested."

Sojourner began making its rounds on the Martian surface, visiting a rock named Barnacle Bill and moving on to a sizable rock named Yogi, where Sojourner made a misstep. Trying to get close to Yogi, it pushed against the rock, and one of its wheels climbed part way up the front of it. That meant Sojourner's chemical analysis instrument—the alpha-proton x-ray spectrometer, or APXS—couldn't be placed against the rock. The rover later backed away enough to use the instrument, and it continued roaming, testing, and exploring.

Matt Golombek was delighted with the results. He had spent his entire career studying pictures from the Viking flight to Mars in 1976. Although Viking was not a geology mission, the camera images provided clues to Martian geology. But those pictures were now more than 20 years old, and Golombek was aching for new data. "The rover is a one-foot geologist, and that little instrument on it, the alpha-proton x-ray spectrometer, gives you the chemistry of all the rocks," Golombek said. "Having that mobility to look around the site, you could start doing the kind of first-order field geology that we do out in the field with a hammer. Now we didn't have a hammer—that turned out to be a problem. But the idea was to try to understand what kind of rocks were there—which is called petrology. If you can identify the rock, you can identify how it formed—that's what geologists do. They look at a rock, and by the training they have, they can tell how the rock formed and they can uniquely tell what the environment was."

Viking, with its retractable arm, had tried to pick up little rocks to take a closer look at them, but it failed. "It was able to pick up things that looked like little rocks, and it tried to analyze them, but they all just crumbled. They were just clods of material." Viking had no equipment to perform chemical analyses of rocks, and mere visual inspection of the rocks didn't tell geologists much, Golombek said. "Basically, you can't tell what those rocks are just by looking at their shapes. You have no idea what their composition is or what they're made of or how they got there or much of anything about them." With the rover, Golombek and his colleagues hoped they would get close-up

FOLLOWING PAGES: The rover, near the end of its mission, captured this close-up of the rock called Chimp. Notice the fracture across the rock and the pitted surface texture. The pits may be gas bubbles frozen in the rock as it solidified if the rock is volcanic, or the pits could be chemically etched by surface weathering.

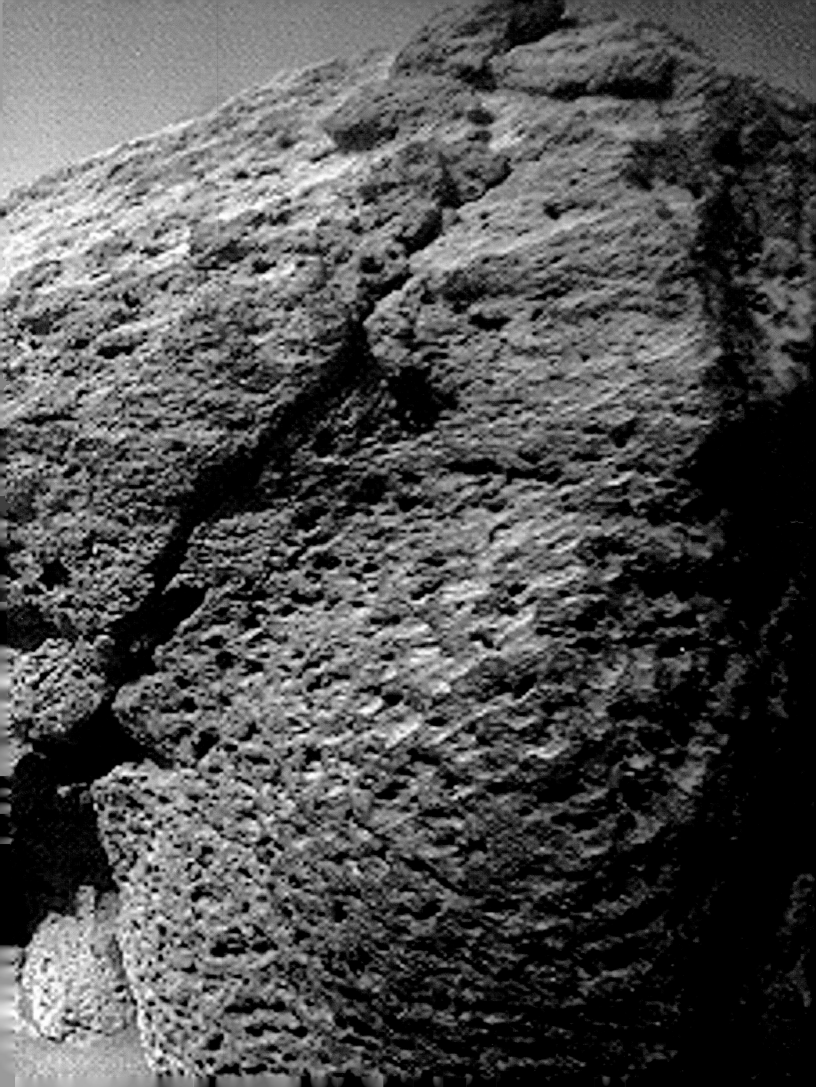

pictures of rocks and be able to perform chemical analyses. That would be combined with analyses of the light reflecting from the rocks as captured by the lander's cameras. "We hoped we'd be able to determine what kinds of rocks were there, and from that infer something about the past history of Mars. That was the most important science that Pathfinder could do."

It didn't take long for Sojourner to surprise and delight the science team. The first rock it encountered, Barnacle Bill, was nothing like what the scientists had expected. It looked like a volcanic rock, because it has small holes in it that looked like gas bubbles created as lava was cooling. That should have made the rock a basalt, like lava flows on Earth or the moon, and like the 12 Martian meteorites that have been found on Earth. "We got the chemistry of that rock back, the amount of silica was way, way higher than in anything that could be called a basalt," Golombek said. "We went to Mars and the first thing we measured was completely unexpected."

If it was indeed volcanic, then Golombek knew it probably had resided in an underground magma chamber or somewhere else in the planet's crust. That leaves time for iron, magnesium, and other heavy elements to settle out of the molten rock, leaving rock high in silica. On Earth, such rocks are called andesites—named after the Andes Mountains, which are made out of high-silica rocks. Barnacle Bill was a distant cousin of the Andes Mountains of South America.

Sojourner continued its explorations and found some rocks high in silica and others with far less. They were covered with varying amounts of red Martian dust, which was lower in silica and higher in sulfur. "One possibility is that these rocks are coated with various amounts of this bright red dust, which would give you this variation in the composition. So they could all be fairly high silica rocks, and if they're volcanic they'd all be this kind of andesite," Golombek said.

Soon, another interesting feature appeared. "We found multiple small, rounded pebbles and cobbles on the surface. These were found in places where we didn't expect them. In one case we rolled over one and it showed up through the dust." One larger rock

seemed to have some of these rounded pebbles on its surface. "The idea is that maybe these aren't all volcanic rocks; maybe some of these are something different," Golombek said. "And then there are these pockets, or what we call sockets, where it looks like one of those pebbles had been—and it fell out." Such a rock, studded with pebbles, is what is known on Earth as a conglomerate. And conglomerates are formed on Earth where rivers have been running for a long time. "With the long period of running water, you can round these pebbles and cobbles and then deposit them into a sand-and-clay matrix," Golombek said. The matrix cements the pebbles together, and the result is a conglomerate.

"That has tremendous implications, if these really are conglomerates," he said. Scientists can't yet be sure that's what they are, but they don't have another good explanation. "We started looking at other rocks, and we started seeing these pebbles and conglomerates all over the place. The only way we know how to form these on Earth is by a long period of water running across the surface to round these rocks and deposit them." That is why these rocks—if they are indeed conglomerates—would have such important implications for the history of Mars. "That argues strongly that liquid water was stable with the atmosphere," and that it remained stable on the surface for a long time. "You need millennia to do this job. And that implies that the climate was much warmer and wetter whenever these rocks formed."

The discovery of what seem to be conglomerates is one of several new pieces of evidence that suggest that Mars was once warmer and wetter than scientists had suspected. The significance of that, in turn, is that it bears on that most intriguing of all questions about Mars: It suggests conditions were right for the formation of life. "This is not the first data to suggest this," Golombek said. "But the way science works is you stack up the evidence on one side of an issue and on the other side of the issue. And I think this—along with some of the other results we have—is suggesting that things were warmer and wetter on Mars."

Golombek sketched out the lines of the argument.

The heavily cratered surface of Mars, he explained, is about 3.5 billion years old. The oldest fossils on Earth are about 3.6 billion years old, and the oldest rocks found on Earth—about 3.9 billion years old—show chemical evidence of life. That is only about 700 million years after the formation of the solar system. The Martian surface, then, records a time when life might have arisen. And the surface is full of old riverbeds and basins that look like the remnants of ancient lakes.

Yet another curious piece of evidence from Pathfinder also supports the liquid-water hypothesis. Pathfinder carried with it several small magnets intended to pull out of the air any magnetic dust that might be present. As the lander sat on the surface, dust did, indeed, accumulate on the magnets. And the color of the dust is red—suggesting that all of the red dust on Mars, the iron-rich rust-red dust that gives it its color, is magnetic. "That's completely unlike anything on Earth, where magnetic sands and dust are very unusual," Golombek said. "So what's going on?"

The theory is that these are tiny particles, probably made up of some sort of clay, that have bonded with a mineral called maghemite—which is strongly magnetic. Maghemite forms only under certain unusual conditions. The most likely explanation is that iron was leached out of the Martian crust by water, and the maghemite was freeze-dried on to little particles of clay once it reached the surface. "It's extraordinary, you almost can't believe it," Golombek said. But the presence of the magnetic dust particles argues that it happened—and that, again, means there likely was a period when water was present. "And it sure helps if things were a whole lot warmer. This is suggesting that water was coursing through there, and it was liquid." Again, the evidence was showing that Mars was once warmer, wetter and more Earthlike.

A third piece of evidence comes from the markings on some of the rocks examined by Sojourner. Some of the rocks are fluted—that is, they are chipped and grooved in a particular direction. On Earth, a marking of this kind is called a ventifact, Golombek explained. It occurs when sand-size particles, less than a millimeter in diameter, are pushed around by the wind, which is strong enough to pick them up and carry them for a bit, before they fall back to the ground and are picked up again. Over a period of time, when these sand-size particles hit rocks, they produce a fluted pattern. "As soon as we saw it on Mars, we said: sand. But we hadn't seen any sand at the landing site." Sojourner continued to explore, however, eventually moving through and beyond a particularly rocky area scientists named the Rock Garden. "Interestingly enough, when we looked a little farther behind the Rock Garden, we found, in the trough behind a ridge, a gorgeous sand dune. This form typically occurs only when you have sand-size particles in this little hopping motion. The wind collects these sand-size particles together and creates these dune forms you see on the surface."

Sand dunes hadn't been seen from orbit, but the discovery of the fluting and the small sand dune raised the possibility that sand was quite common on Mars. And this sand was light in color—meaning that it could be high in silica, just like beach sand on Earth, not low in silica like the SNC meteorites. And sand on Earth is formed by running water. "We don't *know* that that's the case on Mars, but the sand dune on Mars certainly is consistent with the idea of liquid water flowing around and creating, by abrasion, sand-size particles, which are then moved around by the wind," Golombek said.

Golombek summed up the argument: The discovery of what look like conglomerates is fairly strong evidence of a warmer and wetter Mars. The magnetic dust is nearly as strong a piece of evidence. And the fluting and discovery of the sand dune is a reasonable piece of evidence, too. "Then take the idea that you have andesites, or high-silica rocks. If those high-silica rocks are indicative of what the crust of Mars is like, and if you had all this water, then Mars is looking like an Earthlike planet. It's looking completely different from the more primitive and smaller terrestrial planets," Golombek said.

"That's good news if you think Mars is a place where you'd want to go to explore for life. But the key here is liquid water. If you don't have liquid

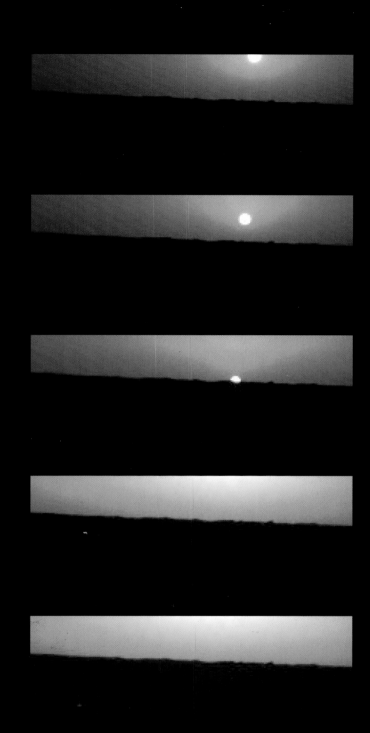

By studying the patterns of solar illumination imaged by Pathfinder at various times of the day, scientists learn about the size, shape, color, and distribution of dust in the atmosphere. The false-color sequence above shows the sun illuminating the morning sky from below the horizon. As the sun rises, the pattern of illumination changes from a broad arc to a more circular pattern over the sun. In the true-color image at left, the sun sets in a pale gray sky.

water, you're not going to have life. You can deal with ultraviolet radiation, with high salinities, with all kinds of bizarre environments. You can have life. But they all have to have liquid water."

Liquid water is, of course, not stable on Mars now. Any water that made its way to the surface would freeze or evaporate, or perhaps occasionally appear as a light blanket of frost. Even if Mars were warmer, its atmospheric pressure is too low to allow water to remain liquid. On a warmer Mars, liquid water would become water vapor unless the atmospheric pressure were greater. But was Mars different billions of years ago? "If it was similar to the Earth, if liquid water was stable at the time that life got started on Earth, then did life form there? If it did, what happened to it? And if it didn't, why not? That's what makes Mars more compelling than any place else to go and study. Mars is just a little more special," Golombek said.

If Mars was indeed warmer and wetter in the distant past, as Pathfinder's scientific results suggest, then what happened? How did it become cold and apparently lifeless? The weaker gravity on Mars might have allowed water to escape. Water vapor can be split by the sun's ultraviolet rays into hydrogen and oxygen, and it's possible that in the weak gravitational field of Mars, the water or hydrogen escaped into space. The Martian atmosphere could have contained more carbon dioxide in the past, but if so, where did it go? On Earth, carbon dioxide is removed from the atmosphere by plants. "On Mars, the carbon dioxide would have to have been taken up in rocks on the surface, and there is little evidence for that," Golombek said.

Perhaps the water is simply hidden underground. "If you believe Mars is a water-rich planet, and it's just too cold now and a little too dry, then you've got to put the water somewhere. And the available evidence is that quite a bit of it could be frozen as ground ice or permafrost underneath the surface," Golombek said. Water ice is visible in the south polar cap, and there is some water vapor in the Martian atmosphere. "Mars could have an enormous amount of subsurface water—hundreds of meters, if not kilometers, in thickness. Mars could be in fact a very water-rich planet."

Pathfinder did return other important scientific results in addition to its geological findings, including those from one experiment that hadn't been planned. The science team had been formed, and planning was well under way when Golombek went to a scientific workshop and realized that Pathfinder could fill in one of the most important outstanding gaps in knowledge about Mars. No one had been able to determine with any assurance whether Mars had a metallic core, and if so, how big that core might be. Golombek realized that the radio signals from Pathfinder would allow its position to be measured within a hundred meters. The signals would also provide very precise information on the rotation of the lander as Mars rotated beneath it. By comparing those figures with similar figures from Viking more than 20 years earlier, the science team could determine exactly how much Mars had wobbled on its axis during the intervening two decades. That wobble would allow the researchers to calculate the density of Mars at various depths—and that would indicate the presence or absence of a planetary core.

"We now know the core can't be any smaller than 800 miles in radius, and we know it can't be much larger than about 1200 miles. That wasn't known, and that came as a freebie—we developed it partway through the mission," Golombek said. "That's a very exciting result—to actually be able to measure how the planet is divided up internally. We have a measure, for the first time, of how big that central metallic part could be."

Pathfinder also provided an interesting look at Martian weather. "We saw the atmosphere in ways we'd never seen it before," Golombek said. In the early mornings, bright white clouds would often appear in the sky in the direction of the rising sun. Some of the clouds had a bluish tint, suggesting the clouds were almost certainly made of water ice. A little later in the morning, the clouds would gradually disappear, as the ice became water vapor.

Pathfinder also found sudden, extreme temperature changes on the surface. Little eddies would pass by that would produce an immediate temperature change of 30°F. And the temperatures a few feet off of the surface were far lower than temperatures right at the surface. "If you were standing on Mars without a space suit—which of course you couldn't do—the difference in temperature between your feet and your nose would be about 40°F," Golombek said. "That's just amazing." You would also see dust devils sweeping by—"that's a twister without the cows," Golombek joked. "We measured them repeatedly, probably every other day or so. We never measured that from the surface with Viking."

Buried within the wealth of data from Pathfinder are many more interesting findings, including some yet to be sifted out and identified. But Pathfinder has already changed Earth's view of Mars. "These results stand out as having significant impact on how we're viewing Mars—the planet, the atmosphere, and the dynamics." All told, Golombek said, "Pathfinder cost less than the movie Titanic, and it had a far happier ending."

Donna Shirley, the director of the Mars Exploration Program, said, "We showed we can land on Mars for $250 million. We can rove on Mars—the rover worked. Before the mission, scientists would sit around and laugh when we proposed this 20-pound rover the size of a computer terminal. They would say, 'you can't do any science with that stupid little thing.' And by George, we can. The makeup of the rocks was possible only because of the rover. You couldn't tell that from the lander, even though the lander had a much better camera. The rover just had these two little cheap Kodak CCD cameras, each about as big around as the end of your little finger. But it was able to get close—and it got great data."

Pathfinder also taught JPL scientists and engineers what it's like to operate on the surface of Mars. "It's very hard, because things aren't the same all the time," said Shirley. "If you're in orbit, things are the same all the time. But if you're on the surface, a dust storm can come up, and the amount of solar energy can change. Or the temperature can change on you, or just about anything can happen. Particularly if you're roving." Some of the tricks used to guide the

rover didn't work very well. The gyrocompasses that recorded its position would drift, and the rover wouldn't know where it was. The wheels would slip, so recording the number of wheel revolutions did not provide an accurate indication of how far the rover had traveled. The rover team is now taking its experience to a mission scheduled for 2003, which will put another rover on the surface, according to Shirley.

Pathfinder also showed that solar power is difficult to use on Mars. The rover operated on about eight watts of power—about as much as a night light. And the lander ran on a few hundred watts of power—less than a hair dryer. But supplying power, even in such small amounts, proved to be difficult. "Batteries that can operate at low temperatures are key," Shirley said. "The rover's batteries just died. We knew they would do that because they were primary batteries, like flashlight batteries—they were not rechargeable. We knew the lander's batteries were going to die, too. They could be recharged but they weren't really designed to be recharged a lot."

Problems with the batteries might have been responsible, ultimately, for the sad end of the Pathfinder mission. The mission was originally scheduled to last a month, but the lander and the rover continued to operate for 83 days—nearly three times what had been asked of them. Even so, when the end came, it was a difficult emotional moment for the Pathfinder team.

Jennifer Harris was the flight director on September 27, 1997. Richard Cook had just left on vacation, leaving Jennifer in charge. Pathfinder, well beyond its normal lifetime, had been working well, continuing to provide data. Harris and her team would send up commands for five or six days of operation. Each morning, the lander would wake up, charge the batteries if it had to, and start taking images. Later in the day, it would point its high-gain antenna at Earth and send the images home. Harris would do a status check, making sure everything was working well on the lander. And the lander would send a picture of the rover at the end of each day to show whether it had gone where it was supposed to. On September 27—sol

Close-up images by the rover cameras helped scientists see detail in the rocks and surface that could not be resolved from the lander-mounted camera. The rock Moe (above, left) shows a grooved surface, which indicates a ventifact (a rock grooved and scalloped by sand-size particles carried by the wind). When such particles hit rock at high speed, they produce this characteristic weathering pattern. Rover images revealed crescent-shaped sand dunes, such as this light-colored barchan (below), that had been hidden from the lander in the trough behind the Rock Garden. The sand hops up the face of the dune toward the rover and then cascades down the opposing face. The wind blows in the direction of the crescent "tails." On the Earth, sand is produced by fluvial processes in which water flows across the surface, breaking the rocks down into smaller particles. In another rover image (above, right), the rock Shark rises behind a smaller rock called Prince Charming. The surface of Prince Charming has small protrusions that look like rounded pebbles, as well as reflective hemispheric sockets, where pebbles may have been. This rock could be a conglomerate, formed by liquid water flowing across the surface for a long period of time, rounding the cobbles and pebbles and then depositing them in a sand and clay matrix. If this rock is a conglomerate, it strongly argues for a warmer and wetter Martian past during which liquid water was stable on the planet. These observations suggest that early Mars may have been similar to the early Earth. Life started on Earth soon after it formed. Therefore, early Mars also may have had an environment conducive to life.

84—the Deep Space Network didn't get the signal.

The best guess was that the lander's battery had failed. That was expected; the battery was not designed to be recharged indefinitely. And it had already lasted three times its planned lifetime. But a plan was in place to operate the spacecraft without the battery, using only solar power.

The power experts were called in for a consultation. "Batteries are like a black science," Harris said afterwards. "You ask the battery guys what happened, and one says one thing and one says another. Nobody knows exactly how the battery will behave." The team devised scenarios to explain what might be happening. One possibility was that the battery failure had thrown off Pathfinder's clock, and that signals were being sent to Earth, but at the wrong time. Yet another possibility was that the spacecraft—confused about the time—was cycling through a series of past commands ending with a command to "power off." Ground control had no opportunity to break that cycle. If that were true, the spacecraft would turn itself on every morning, run through the sequence, and shut itself down. This was referred to by Pathfinder engineers as the Kevorkian syndrome.

By sol 87, three days after the failure to hear from the spacecraft, Pathfinder needed special help from the Deep Space Network's 70-meter antennas, its largest and most sensitive antennas, which were then busy communicating with the Galileo spacecraft, a Jupiter mission then touring Jupiter's moons. So Harris officially declared a spacecraft emergency, automatically giving Pathfinder the highest priority in the use of the Deep Space Network. The Deep Space Network's 70-meter antennas were now at Harris's disposal.

Communication with spacecraft is done by sweeping through a narrow range of frequencies—in essence, checking a lot of different channels in a particular radio band—until a signal is heard. Receivers then lock on to that frequency. The frequency, or channel, on which spacecraft communicate can vary with the temperature of the spacecraft, among other things. The transmitter on the spacecraft could have been off for three days at this point,

and it would be getting cold—and the frequency on which it was transmitting had probably drifted. Harris's plan was to send a command to the auxiliary transmitter to disconnect the battery and switch the spacecraft to solar-power operation. "We would do a sweep, try to turn on the transmitter, and wait. The light time was 11 minutes, so we had a 22-minute wait before we could see anything," Harris said.

It was 2 a.m. when these attempts were being made. For days, Harris and her team had been up all night trying to contact the spacecraft, and busy during the day trying to retain control of the Deep Space Network. "We were there all the time. We were really hoping to see the signal, because we knew if we didn't it was pretty serious." Galileo mission controllers were desperately trying to get back control of the 70-meter antennas. "They put up an unbelievable protest—and rudely so," Harris said. "The second day they called us and told us to give up. I wasn't going to give up." Harris, project manager Brian Muirhead, and a half dozen others went into a conference room to try to clear their heads and begin again to rethink the problem. "It was kind of the 'end of Pathfinder' meeting," Harris recalled. "Nobody wanted to be there....It was three in the morning, and we were all extremely depressed. And on the voice boxes we hear this guy from the Deep Space Network station in Canberra say, 'I see a signal.' You never saw six people move so fast."

The team immediately resumed efforts to regain contact with the spacecraft, but the efforts failed. Richard Cook returned from vacation, and the team once again tried every imaginable scenario. Even though Pathfinder had survived for three times its planned lifetime, even though it had provided a masterful engineering demonstration and returned a bounty of scientific information, the end came as a blow. "This thing was people's lives for a long time," Harris said.

Several months later, she still felt the sting of the disappointment. It was difficult, she said, "to move on from something that was so cool, and so much fun, and so exciting."

The team never determined exactly what had gone

Naming the Rocks

Naming Martian planetary features often proves serious business—but not so with the naming of the rocks. As Pathfinder's Sojourner rover wheeled slowly across the Martian surface, it encountered Barnacle Bill, Yogi, Pop-Tart, Shark, Half Dome, Moe, Stimpy, Cabbage Patch, and Chimp, among many others (left, both images). Pathfinder's chief scientist, Matthew Golombek, explained that the Pathfinder scientists chose to use the names of cartoon characters and other easy-going designations out of convenience. "The ground rules were pretty simple," says Golombek. "You couldn't name anything after a person, a place, a family member, or a dog. And it had to be lighthearted." The system for deciding upon names merely involved writing a name on on a sticky note and placing it on a 15-foot panorama of the landing site that had been taped to the wall. Every few days a "name czar" selected from the team went through the proposals and decided which ones would become permanent. "As soon as we landed, the press asked if we'd named the rocks," Golombek recalled. "We hadn't even looked at anything yet." But the naming soon began. The idea of using cartoon characters for most of the rocks arose spontaneously. But a few other names came up as appropriate, too. "One of them is called Broken Wall. It was named by one of the Germans after the dismantling of the Berlin Wall," Golombek said.

On Mermaid Dune on sol 30 the rover faces the lander. Mermaid Dune is covered by dark gray soil; the image also shows dark rocks, dust, and dark red soil exposed in rover wheel tracks. The rover has dug its wheels into the surface to determine the mechanical properties of these deposits. Rover images also helped distinguish the size of the particles.

wrong. A leading theory is that the spacecraft receiver had become too cold and, under the strain of temperatures for which it hadn't been designed, it simply broke. At the time, the team refused to declare the mission officially over. "Nobody wanted it to be over," Harris said. "We had a final press conference on November 4 and said we've gone into our contingency mission and were going to try to communicate every couple of weeks."

Even so, it was time for the team to move on. "This was a great team," Harris said. For some weeks before the end, the mission had become routine, but the crisis renewed the doggedness and determination that had characterized the team from the start. "The team came back together for those last two or three weeks. It was Pathfinder again. Everybody in the room, they were people you could learn a lot from. Everybody would come in at about one in the morning, and we'd run everything through the test bed and make sure everything was ready. We would work until about seven. Then we would go and have breakfast in the cafeteria. Then we would fight with Galileo. We didn't sleep for about the first three days. I think I aged 10 years in two weeks, but I learned so much." Harris, the flight director for Pathfinder's final days, was 29 years old.

Engineers continued to make occasional attempts to contact Pathfinder during the next few months. On March 10, 1998, the Pathfinder team announced it would make its final attempt to contact the spacecraft. Flight controllers spent nearly four hours commanding the lander to turn on its transmitter and waiting for a response. At 1:21 p.m. Pacific Standard Time, the Pathfinder mission was over. Brian Muirhead said no further attempts would be made to communicate with Pathfinder.

What about Sojourner, the one-foot geologist? Every day during the mission, flight controllers would play a wake-up song for Sojourner. During the crisis, after trying everything to regain communications with the spacecraft, Harris thought a wake-up song might do the trick. "I brought in my Simon and Garfunkel tape with 'Wake Up Little Susie.' We had run out of science solutions, and I thought maybe this would do it. So I hit the tape, and it played two measures of 'Wake Up Little Susie,' and then something made it flip over to the other side, and it played 'Bridge Over Troubled Waters.'" Each day during the crisis, the rover team would come into mission control with its plans for that day's activity, hoping that the spacecraft was back in operation. "That was one of the hardest things, with the rover guys coming in every day saying, 'O.K. this is our plan.' And we were saying, 'Can you give us another few days?'"

The rover had been designed so that if it didn't hear from the lander in seven days, it would conclude that it was out of range, and tiny Sojourner would head back to the lander—to its home base. But it also knew not to get too close to the lander—a "keep-out zone" had been established so the rover would not crash into the lander and damage it. When the spacecraft failed, Sojourner was in the Rock Garden. When it didn't hear a signal, it could have turned and headed back over the rocky surface to the keep-out zone. Eventually, it would have run out of power or become stuck on a rock. Until then, however, it would have remained just out of reach of the lander, slowly bumping along, in a small circle, waiting for any word.

Despite its inevitable ending, Pathfinder had been one of the most successful planetary missions in NASA's history. America's space exploration program was back on track. Mars had become a friendlier and more familiar place. And the future of Mars exploration was brighter than ever.

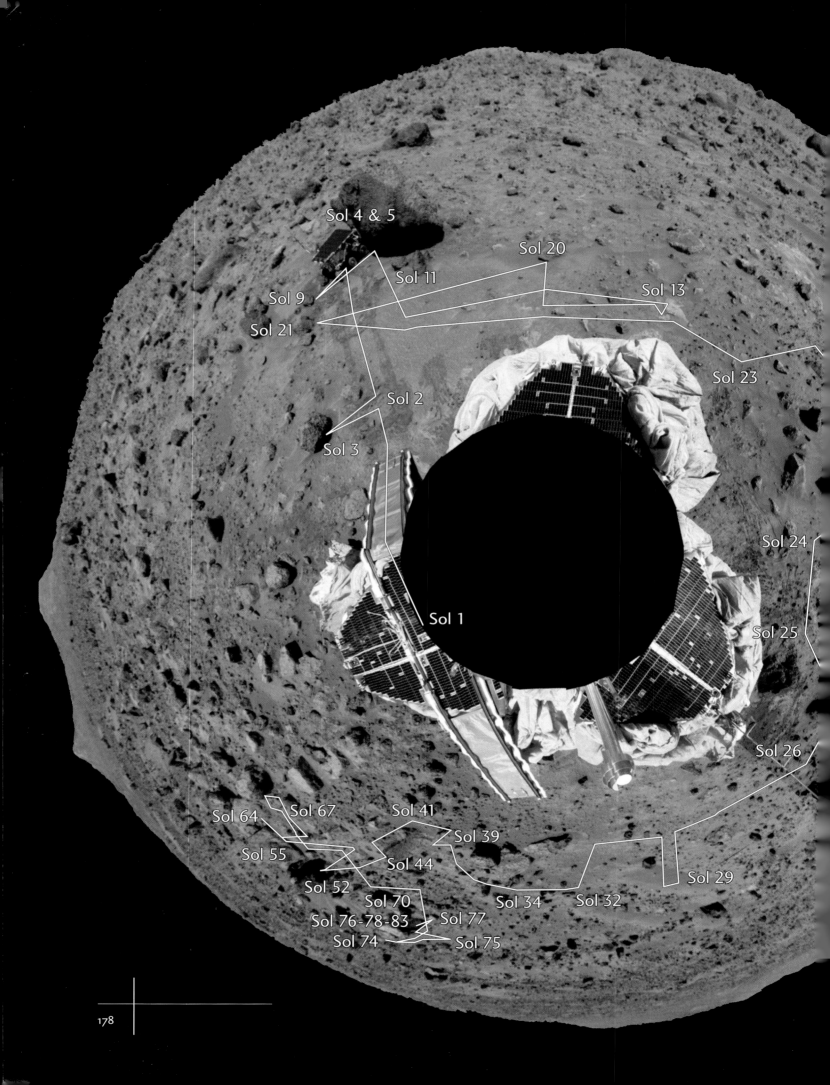

Sol 4 & 5

Sol 20

Sol 11

Sol 9

Sol 13

Sol 21

Sol 23

Sol 2

Sol 3

Sol 24

Sol 25

Sol 1

Sol 26

Sol 64

Sol 67

Sol 41

Sol 39

Sol 55

Sol 44

Sol 52

Sol 29

Sol 70

Sol 34

Sol 32

Sol 76-78-83

Sol 77

Sol 74

Sol 75

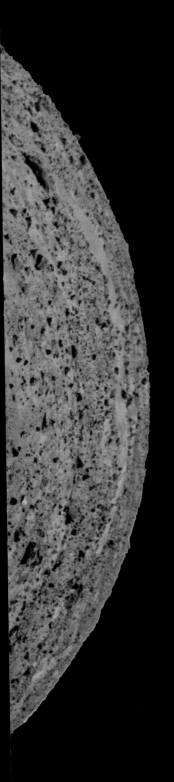

Mosaic depicts a perspective bird's-eye view of the Pathfinder lander, the landing site, and the traverse of the rover marked by sol (Mars day, which is 24 hours and 37 minutes, with sol 1 being the day of landing). The lander is in the center (the blank circle), with the three triangular petals protruding. Air bag material is seen beneath the petals, which are covered by solar panels for generating power for the lander. The thin mast at the end of the petal to the southeast (north is up) is the meteorology mast for measuring temperature at three heights as well as wind speed and direction. The metallic cylinder is the low-gain antenna, capable of returning data directly to Earth at up to a few hundred bits per second. The rover petal has the two ramps, with only the rear ramp touching the ground.

Consequently, the rover exited the rear ramp on sol 2 and measured the composition of Barnacle Bill on the night after sol 3. The rover then went toward Yogi, which it measured after performing a number of soil mechanics experiments. Next the rover sampled a variety of soil types, distinguished mostly by color near Scooby Doo (sol 13). After testing out the autonomous navigation system in this region, the rover began a clockwise drive around the lander to get to the area known as the Rock Garden (sols 52-67), which could not be reached from the other direction. This area ranked as a high priority to the science team because it has large rocks with steep faces that did not appear covered by dust, so that chemical analyses could be made of the rock instead of the dust.

n the spring and summer of 1995, in a NASA laboratory in Houston where the first moonrocks were studied, two researchers were secretly working to try to confirm and interpret the most explosive scientific data they had ever seen. They decided to guard their findings carefully until they had tried every possible way to disprove them. Theirs was the scientific find of the century, or even more—if it proved to be true, one of the most profound scientific discoveries of all time. "We spent a year or a year and a half trying to debunk ourselves," said one of the researchers, Everett K. Gibson, Jr. He and David S. McKay had been colleagues at the Johnson Space Center in Houston from the time they had worked together during the Apollo missions, analyzing rocks brought back from the moon. McKay had helped train the lunar

In this artist's rendering, a vast ocean fills the northern lowlands and the canyons of Valles Marineris. Some scientists find indications that Mars could have had such an ocean in times past.

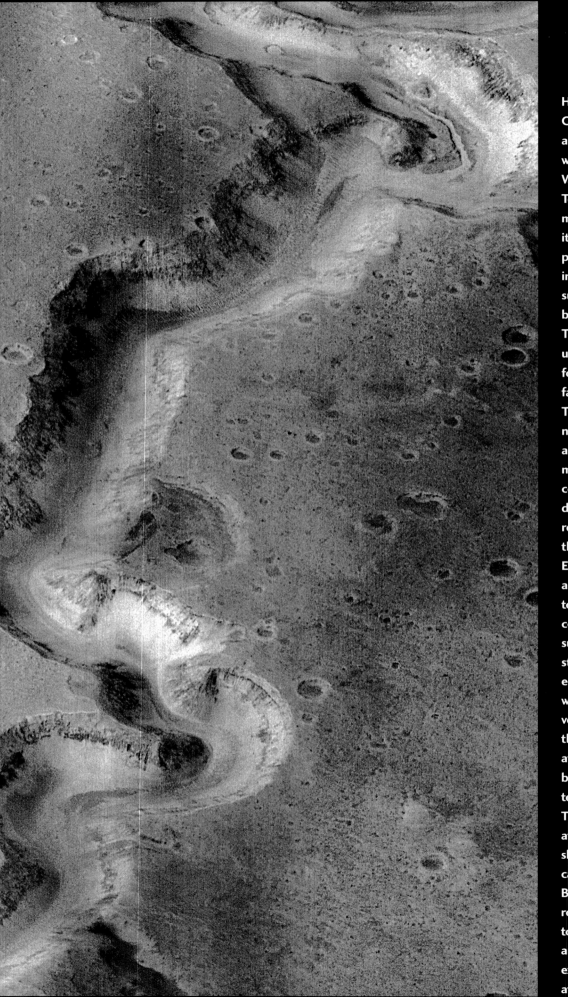

High-resolution Mars Global Surveyor image about 6 miles (10 km) wide reveals Nanedi Valles in the Xanthe Terra region. About 1.5 miles (2.5 km) across, it cuts into a cratered plateau, revealing layers in its upper walls that suggest deposition by water or volcano. The steepness of the upper cliffs indicates formation by debris falling into the canyon. The lower canyon has a narrow central channel about 656 feet (200 m) wide, suggesting continual flow carrying debris. The meanders remarkably mirror those found in rivers on Earth, with tight curves and an oxbow near the top of the image. Such conditions would suggest that water was stable in the Martian environment, which would have required a very different climate than exists now. The atmosphere would have been denser and the temperature higher. The carbon of that atmosphere would show up today in carbon-bearing rocks. But not enough such rocks have been found to correlate with the amount of carbon expected in a denser atmosphere.

astronauts. But nothing discovered in the lunar samples could compare with what they now held in their hands. "This could be a breathtaking conclusion," NASA Administrator Daniel Goldin said when he learned of it. He personally interrogated the scientists for two and a half hours before making up his mind about their claim. When he finished debriefing them, he asked NASA public affairs officials to call a press conference.

McKay and Gibson had been immersed in the examination of a four-and-a-quarter-pound, gray-green rock that had been found in the ice in Antarctica in 1984. The rock, about the size of a large potato, is about 4.5 billion years old. It fell to Earth as a meteorite 13,000 years ago. The nature of the rock was not determined until 1993, when researchers discovered that it was one of the rarest of all known meteorites—a meteorite from Mars, blasted from the Martian surface by the impact of an asteroid, perhaps, or by some other cataclysmic event. Only 12 such Martian meteorites are known. Referred to as the Shergotty-Nakhla-Chassigny, or "SNC," meteorites (pronounced "snick"), they were identified as pieces of Mars based on their chemical composition, which differs from that of the Earth and the moon. The SNC meteorites provide a unique opportunity to study Martian geology, at least until NASA arranges a mission to collect rocks on Mars and bring them back to Earth.

The rock McKay and Gibson were studying left Mars 16 million years ago and then drifted through the solar system until it was captured by Earth's gravity and fell on the Antarctic ice. There it rested through the millennia, probably buried in compressed snow and later exposed. In 1984, a geologist spotted it, bagged it, and sent it to the United States for study. The rock was found in the Allan Hills of Antarctica. It was the first meteorite from the 1984 expedition to be processed in the lab, so it was catalogued as ALH (for Allan Hills) 84001.

McKay and Gibson and their colleagues began studying the rock in 1994, shortly after it was recognized to be a piece of Mars. They had studied it for about a year when they realized that they were on the verge of what could be a colossal discovery. They spent another year or more doing further studies to be sure that they were right. "This is the result of two-and-a-half years of very intensive, meticulous and difficult detective work," Wesley Huntress, Jr., NASA's associate administrator for space science in Washington, said when the discovery was announced.

After checking and rechecking their findings, McKay and Gibson submitted their research to *Science* in mid-1996. The editors of *Science* kept the discovery confidential as the report was being reviewed and prepared for publication. The report was scheduled to be published in the August 16 issue, but in the final few days before the journal appeared, word of the discovery began to leak. The press conference that Goldin had asked for was held in Washington on August 7. There, McKay and Gibson made their dramatic announcement: They reported that they had found a chain of evidence suggesting the presence of life on Mars.

Goldin opened the press conference with words of caution. "The results today are not conclusive," he said. "We are not here to establish, as in a courtroom, beyond a shadow of a doubt, that life existed on Mars. Instead, he said, the researchers had found "compelling clues" suggesting that "life might have existed, at some point, on Mars." Goldin called on other researchers to investigate and evaluate the findings. The discovery "is certain to create lively scientific debate and controversy," he said, and "that's what makes American science and world science great." Goldin announced that samples of the rock would be made available to other investigators. "We want to take time to do this, and if it takes a year, or two years, so be it."

It is doubtful that Goldin foresaw what was to come. The announcement set off a furious scientific argument. McKay and Gibson suffered withering attacks at the hands of other scientists. The two met each criticism with equally fierce arguments defending their work. And as the two-year anniversary of the announcement approached, there seemed to be no end to the debate.

At that first press conference, McKay explained

that there were various lines of evidence in the rock to suggest the presence of life on Mars in ancient times. Much of the evidence dealt with the discovery in the meteorite of tiny globules of calcium carbonate—the main constituent of limestone. McKay said those carbonate globules could have a biological origin. The rock was also found to contain organic material that McKay and Everett believe came from Mars. And ALH84001 contains what look like fossils of extremely small organisms of some kind. "This is perhaps the most controversial part of our presentation," McKay said.

While that might have been the most controversial aspect of the discovery for scientists, it was the easiest for the public to grasp. As he showed slides of the tiny structures at the press conference, McKay said, "The features that you see may be any number of things. For example, they could be dried-up parts of clay, or they could be microfossils from Antarctica. Or microfossils from Mars. It is our interpretation—the one that we favor—that these are, in fact, microfossil forms from Mars." He reiterated that this was only an interpretation of the findings—that there was no independent confirmation that these were fossils from Mars. To nonspecialists, the tiny structures McKay was displaying looked very much like bacteria or tiny worms. That would not do for scientists, though. McKay himself warned about "the pitfalls of identifying such things based on appearance alone."

To anticipate the criticism that the discovery was likely to provoke, NASA took the unusual step of inviting J. William Schopf of UCLA to the press conference. Schopf was not connected with the discovery, and he did not share all the views of McKay and Gibson and their colleagues. Goldin said Schopf was invited to provide a "point-counterpoint" on the evidence at hand. Schopf, who has spent 30 years studying the origin of life on Earth, began his presentation by noting that, as Carl Sagan once said, "extraordinary claims require extraordinary evidence."

"I happen to regard the claim of life on Mars, present or past, as an extraordinary claim," Schopf said. "And I think it is right for us to require extraordinary

evidence in support of that claim." He began by showing a picture of the oldest evidence of life on Earth—microscopic fossils from 3.465 billion years ago. He noted that fossil cells are clearly visible in the samples, and the cell walls are made of organic matter. And he pointed out that a small piece of one of the fossils—which are among the smallest fossils on Earth—is 100 times the size of the fossil-like objects in the Martian meteorite. Could objects so much smaller than any fossils known on Earth actually be fossils?

Schopf raised other questions about McKay and Gibson's momentous announcement. He noted, for example, that there was disagreement over when and how the globules of carbonates in the meteorite formed. Some argued that the carbonates formed at very high temperatures or during the impact that blasted the rock from Mars. If that was true, then the carbonates were likely the result of some physical process, because the impact or the high temperatures would have destroyed any life in the rock. He acknowledged that ALH84001 contained organic molecules, but he noted that those molecules could have come from nonliving sources. There are many ways organic chemicals can be formed in the absence of life. The fossil-like structures might be what are called pseudo-fossils, made of minerals, not from the remnants of living organisms, he said. At the end of his presentations, he said he did not believe the researchers had found conclusive evidence of life. "I think they're pointing in the right direction," he said, but "additional work needs to be done before we can have firm confidence...of life on Mars."

The report by McKay, Gibson, and their colleagues appeared in Science under the title: "Search for Past Life on Mars: Possible Relic Biogenic Activity in Martian Meteorite ALH84001." They had enlisted Richard N. Zare, a Stanford University chemist whose expertise lies in the identification of organic compounds. He determined that the organic compounds in the meteorite—substances called polycyclic aromatic hydrocarbons, or PAHs—originated in the rock and were associated with the carbonate globules. After running through the details of the data, and explaining what they had done to avoid contaminating the meteorite in the laboratory, the researchers listed various lines of evidence consistent with their claim that the rock held evidence of past life on Mars. The biological activity—if that's what it was—occurred where fluid had penetrated cracks in the rock. The carbonate globules formed later than the rock itself. The rock seems to contain fossils of microorganisms. Magnetic particles in the rock could have been formed from chemical reactions in living organisms. And the PAHs Zare had examined were found near or on surfaces rich in carbonate globules.

"None of these observations is in itself conclusive for the existence of past life," they wrote. But "when they are considered collectively, particularly in view of their spatial association, we conclude that they are evidence for primitive life on early Mars." The debate was on.

The size of the microfossils quickly became an issue. While some said they were ten or a hundred times smaller than anything known on Earth, one infectious disease specialist pointed out that they were only five times smaller than the smallest living bacterium on Earth. It is entirely plausible that organisms of that size could have evolved on Mars, he said. McKay and company responded that structures the size of those in the Martian meteorite had been found in copper ore. They called the organisms that created the fossils "nanobacteria." Support arrived from a British team studying another one of the SNC meteorites. The British scientists determined that its organic compounds were the kind that would most likely be produced by microbes, not by chemical processes. Some argued that the organic compounds identified by the NASA researchers were actually from Earth. These critics argued that ALH84001 had been contaminated by organic substances after it reached Earth. In each case, McKay, Everett, Zare, and the others provided rebuttals.

By the time of the Lunar and Planetary Science Conference in Houston in March, 1997, 45 laboratories had applied to NASA to study the meteorite. Two of the main sticking points that emerged as the debate continued were the temperature at which the carbonates had formed, and whether the structures seen in the Martian meteorite were too small to have been living organisms. They could have originated in minerals rather than in living cells, some researchers argued. Another issue that continued to be debated was whether the PAHs originated on Mars or were contaminants picked up on Earth.

McKay and Gibson never shrunk from the debate. In an article in *Scientific American* in December 1997, they laid out the case for and against life as they then saw it. The carbonates in ALH84001 have not been found in any of the 11 other meteorites from Mars, they noted. "Basically, the case for ancient microbial life on Mars is built almost entirely around the globules," the researchers wrote. For one thing, the globules contain tiny iron oxides and iron sulfide grains similar to those produced by bacteria on Earth. When that piece of evidence is combined with the presence of organic carbon molecules and fossil-like structures, the case becomes stronger, they said.

In response to critics who said the carbonates must have formed at very high temperatures, McKay and Gibson also pointed to new evidence that the carbonates formed at temperatures below 100°C, or 212°F—the boiling point of water. That is not too high to eliminate the possibility of life. "The arguments for high-temperature carbonates have not considered everything in the rock," Gibson said. An analysis of the isotopes, or different forms, of carbon in the globules showed that the carbonates in the rock have more carbon 13 than any carbonates found on Earth—"but just the right amount to have come from Mars," the researchers wrote. Although PAHs can arise from nonliving sources, the researchers pointed to evidence that keeps them convinced that the PAHs in the rock are of biological origin, they wrote.

The issue was debated again at the Lunar and Planetary Science Conference in Houston in March 1998. In an interview summing up the conference, Gibson said, "The bottom line is we feel more strongly now about the hypothesis than when we wrote the manuscript."

Dubbed Allan Hills 84001, or ALH 84001, (above) this meteorite has spawned a wide-ranging debate in the scientific community about whether it really harbors evidence for past life on Mars. Believed to have been blasted off the surface of Mars about 16 million years ago by an impact, the 4.5-billion-year-old meteorite arrived at Earth about 13,000 years ago. Its name derives from the ice field in Antarctica where it was found. The cut section shows mafic igneous rock with cracks filled with carbonate globules that have been dated by various research groups at between 1.3 and 3.6 billion years old.

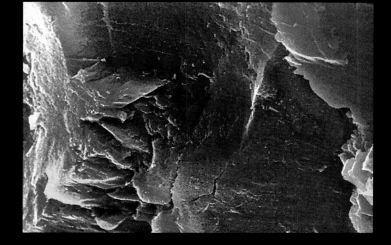

Does the meteorite ALH 84001 exhibit evidence for past life on Mars? A group of scientists argue that primitive bacteria living in the cracks within the rock produced the carbonate globules, magnetite crystals, and organic molecules (polycyclic aromatic hydrocarbons) found there. The orange carbonate globules (right, lower two images) measure a few hundred microns; the magnetite and iron sulfide crystals, ten to a hundred microns. The small wormlike and ropelike features of the meteorite—only 20 to 100 nanometers long—made visible in these scanning electron microscope images (right, upper two) have been proposed as possible fossil bacteria. The structures indeed bear some resemblance in shape to bacteria microfossils on Earth. Other scientists disagree. They have found evidence that the carbonates and magnetic crystals formed in temperatures too hot to support life, that the organic molecules could be contaminants from the Antarctic ice, and that the microfossils are too small to even fit the chemicals necessary to support life. Given the small size of these structures and the inherent difficulty in studying things this minuscule, it may never be possible to say definitively whether life existed in this rock. As a result, one of the most important goals of the entire Mars exploration program is to determine whether or not life started on Mars early in its history.

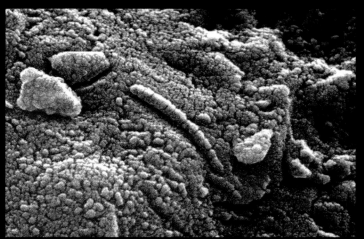

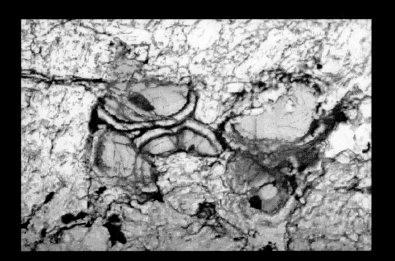

As to whether the rock was contaminated on Earth, Gibson admitted that "every meteorite from Antarctica contains terrestrial amino acids." But when the carbonate globules are examined separately, they show evidence not of terrestrial contamination but of material from Mars, Gibson said. At the 1998 meeting, Kathie Thomas-Keprta, one of Gibson's colleagues, presented evidence in the rock of magnetic material called hexagonal magnetites, which, Gibson said, can be produced only by bacterial processes. The researchers also found evidence of "biofilms," organic material secreted by organisms to protect themselves from hostile environments. And increasing data on the carbon isotopes is showing "that these are in the direction living systems love," he noted. "Our data has stood 19 months," Gibson said. "It has not been refuted. It has not been shown to be wrong. What has happened is people are upset with our hypothesis."

Gibson acknowledged that the scientific debate, while exciting, has also been difficult. In the 14 months following the announcement, Gibson and his colleagues counted more than 500 presentations they had made to scientists and to the press. "We've made every opportunity to go to every meeting we've been able to get travel funds to go to," he said. "We want people to see that all the data is there....People are beginning to realize we're doing solid work."

Gibson compares the initial reaction to a couple of earlier events in the history of science. When plate tectonics was first proposed—the idea that vast plates on the Earth's surface are drifting lazily around the globe—nearly 50 years passed before it was accepted, Gibson noted. It took two decades or so for researchers to accept the idea that the Earth was bombarded by a huge meteorite at the time the dinosaurs disappeared. And it may take years for the question of life on the Mars meteorite to be resolved, he said. "This is a complex scientific problem....We'd like to think this will be resolved when samples come back from Mars."

Gibson's and McKay's meteorite speaks to the possibility of life on Mars eons ago. What about the possibility of life on Mars now? If life existed in ancient times, could it have survived? A couple of

decades ago, many researchers who had studied Mars would have said the odds of life existing now or in the ancient past were very low. But some of them now say the odds have risen. Not because researchers know more about Mars—but because they know more about life on Earth. Living organisms unknown in Viking days have since been found in some of the most extreme, unlikely environments on Earth—raising hopes that life might have once gained a foothold on Mars, and that it might be hanging on even now, perhaps hidden somewhere underground.

In the 1920s, researchers suggested that life on Earth arose from chemical reactions in an ancient Earth atmosphere rich in hydrogen and low in oxygen. The idea was suggested by the Russian biochemist A. I. Oparin, who thought that hydrocarbons and ammonia could have led to the formation of complex organic compounds and eventually to life. A few years later, the biologist J. B. S. Haldane suggested that organic compounds might have accumulated in the oceans, leading to the formation of a "hot dilute soup" that in turn gave rise to life. In 1952, those ideas were picked up by Harold Urey, who realized there was a way to test them. One of Urey's students, Stanley Miller, tried to duplicate Earth's early atmosphere by putting hydrogen, methane, ammonia, and water vapor in a container. He then ran an electric charge through the mixture, simulating the effect of a lightning bolt, and he found four amino acids and other chemicals essential to life. The experiment provided a plausible explanation for how life might have originated on Earth.

In the years since, however, it has become clear that life can form in many other ways. Shortly after Viking's visits to Mars, researchers exploring the ocean floor discovered strange hot vents, where bubbling gases streamed up from beneath the ocean floor. To the astonishment of the researchers who found them, these hellish vents were covered with swarms of worms, crustaceans, and other forms of life unknown anywhere else on Earth. This was life that presumably originated independently of life on the surface.

The discovery of the vents—which are sometimes called "smokers"—raised the idea that life on Earth

The bubbling cauldrons of Wyoming's Yellowstone National Park rank among the most inhospitable environments on Earth. Long thought to be lifeless, the geysers, hot springs, and steaming pools have proved inhabitable by a variety of unusual organisms, sometimes called extremophiles for their "love" of extreme environments. The discovery of microbes in geysers (above), hot springs (right) and hot pools (top) raises hopes that life might exist on Mars.

The Greatest Mars Watcher of Them All

An intellectual wanderer, Percival Lowell worked in his family's textile business in Massachusetts and dabbled in Asian diplomacy before turning to astronomy. He was not a trained astronomer, but his research on Mars so dominated thinking about the red planet that, even now, more than a century later, his influence is still felt. Lowell made substantial contributions to astronomy, but he is best remembered for his increasingly elaborate descriptions of the maze of canals crisscrossing the Martian surface. His work on the canals stands as one of the greatest masterpieces of self-deception in scientific history.

The Lowells, descendants of some of the early Massachusetts colonists, were among Boston's wealthiest and most important families. Percival, born in 1855, had a childhood interest in astronomy, but he chose to study mathematics at Harvard. When he graduated, in 1876, he already had a reputation as one of the most brilliant young men in Boston. He turned down a prestigious invitation to teach mathematics at Harvard and entered the textile business, but soon grew restless. A few years later, he left for Asia, where he pursued the life of a scholar and helped to write a new constitution for Japan. In the 1890's, he renewed his interest in astronomy when he returned to the United States, and decided to expend some of his considerable wealth building an observatory to prepare for the 1894 opposition, when Mars would be only 41 million miles away.

Lowell became convinced that the Arizona Territory would be an excellent spot for astronomical observations. Unable to work out an agreement with Harvard

Percival Lowell observes Mars at the observatory he built in Flagstaff, Arizona. He had little formal training in astronomy, but through his observations of Mars Lowell contributed enormously to knowledge of the planet and stoked public enthusiasm for its exploration. His unyielding adherence to the idea of Martian canals and intelligent life, however, led other astronomers to ostracize him. He submitted many research papers to scientific journals only to have them rejected.

> *"That Mars is inhabited by beings of some sort or other we may consider as certain as it is uncertain what those beings may be."* – **PERCIVAL LOWELL**

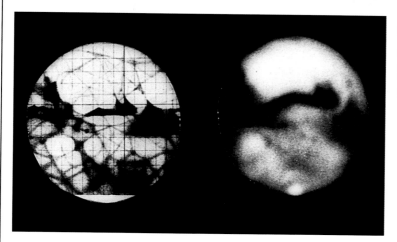

Early drawing of Mars canals (left) as seen by Percival Lowell is juxtaposed with an actual photograph of the red planet.

to sponsor an observatory there, he decided to build it himself, with the help of William H. Pickering and Andrew Douglass, Harvard astronomers who took leaves of absence to go to Arizona. The Lowell Observatory was built on top of a 7,000-foot mesa near Flagstaff. In May, 1894, Lowell and Pickering began their observations, as Mars was growing closer to Earth and rising above the horizon, where it could be more easily seen. Lowell reported that he could indeed see a faint grid on the planet's surface, confirming the existence of the canals that he had by then read so much about.

Based on those observations, he mapped 184 canals—more than twice the number of canali reported by Giovanni Schiaparelli a dozen years earlier. He then turned the force of his estimable personality to the task of spreading word of his discovery. Many scientists disputed his claims, but the public responded eagerly. He wrote a series of articles on Mars for the *Atlantic Monthly* and *Popular Astronomy*, though his submissions to the *Astrophysical Journal* were rejected. He published his canal maps in 1895 in a book entitled *Mars,* the first of his books on Mars.

There is no question that Lowell believed what he saw. Through dint of sheer brilliance and persistence, he found order where none existed. He studied the fuzzy image of Mars in his telescope as though it were a crystal ball, and he spied all kinds of wonders in it. As knowledge of Mars grew over time, belief in Lowell's beguiling canals gradually faded away, as did the idea that Mars might harbor intelligent life. In 1907,

Alfred Russel Wallace, one of the discoverers of the principle of evolution by natural selection, delivered what should have been a knockout blow to Lowell's canal theories. Asked to review one of Lowell's books, he found himself so absorbed in the task that he wrote a book-length essay. He identified an error in Lowell's calculations claiming to show that the temperatures of Mars were comparable to those of southern England. Temperatures must be below freezing, Wallace said, arguing that the planet was likely covered with permafrost. It was also likely to be covered with craters, and any attempt to move water by canals would fail, because the water would evaporate or soak into the soil, Wallace said. "This devastating and largely correct physical analysis was written in Wallace's eighty-fourth year," wrote Carl Sagan. "His conclusion was that life on Mars—by this he meant civil engineers with an interest in hydraulics—was impossible."

That didn't stop popular writers from picking up Lowell's ideas, however. Most notable among them was Edgar Rice Burroughs, whose novels of life on Mars began appearing in 1912. In 1971, Bruce Murray, a planetary scientist at the California Institute of Technology, lamented that "we are all so captive to Edgar Rice Burroughs and Lowell that the observations are going to have to beat us over the head and tell us the answer in spite of ourselves." The 1996 report that a Martian meteorite appeared to contain fossilized evidence of life excited popular enthusiasm that would have cheered Percival Lowell. His legacy continues.

might have originated not in the "dilute soup" of the oceans but perhaps in some of Earth's most hostile environments—near heat and water. While that discovery was being made, molecular biologists were devising ways to create vast family trees to trace the ancestry of today's millions of species. When the family trees are followed to their roots, using RNA as a guide, they point to organisms that live in very hot environments. According to the family tree studies, these organisms are the earliest ancestors of life on Earth. Some of their direct descendants "live in these smokers at the bottom of the ocean, and they live in the geysers in Yellowstone," said Michael H. Carr, a geologist at the U.S. Geological Survey in Menlo Park, California, and one of the deans of Mars science. "People suggested that this may well be where life started on Earth, not in quiet tidal pools, which was the former thought. So when we think about Mars— we know Mars was extremely volcanic early on. We think we know it's water rich. You put water and volcanism together, and you get hydrothermal activity—which is what you have at Yellowstone."

"In recent years, it's become clear that life is pervasive and can live in all sorts of environments," Carr said. One of the most interesting examples comes from research done during the past few years that has shown that microbes live in basalt formations nearly two miles below the Earth's surface. The discovery of life so deep underground came as a surprise. The organisms that have been found there "have a very low metabolic rate," Carr said. "They live independently of the surface and appear to get energy from the interaction of ground water and basalt, which produces a little bit of hydrogen. These organisms can live on this little bit of energy at very low metabolic rate—they may reproduce only once a century."

It's widely thought that life can't exist on the surface of Mars. Conditions are too harsh. But if life did originate on Mars, it might have survived below the surface. "All of these speculations make people a little more optimistic than they were 10 or 15 years ago," Carr said.

Mark Adler of the Jet Propulsion Laboratory said the search for life on Mars is now "centered around environments where there was liquid water and energy. Those include ancient hydrothermal environments, places deep beneath the ground where there is heat from the core of the planet providing the energy for life and for liquefying water—and providing the water itself. The theory is if we can find rocks formed in these ancient hydrothermal environments, maybe we can find evidence, preserved in the rocks, that something once lived there." If, however, Martian life exists nearly two miles below the surface—as it does on Earth—finding it could be impossible. "What we'd really like to find is the Yellowstone Park of Mars," Adler said. "Of course, as soon as the water comes up it will freeze and very quickly sublimate. Water is not stable on the surface, it will go away almost immediately. But if we can get there where it's happening and bring some back and preserve it, it would be very interesting to look and see whether there are any life forms in it."

Adler believes that if life ever existed on Mars, it is probably still there. "On Earth, life always finds a way to survive somehow. It changes and adapts and finds a way to survive," he said. "There are a lot of models that indicate that there could be liquid water deep beneath the surface of Mars, where life could exist today. I would say that if there ever was life on Mars, it's found a way to survive in that environment today. It doesn't mean that we can get to that environment, and it doesn't mean that we will find evidence of it, or even if we do find evidence that we will recognize it right away."

One puzzle remains to be solved about life on Mars and its relationship to life on Earth. The SNC meteorites provide evidence that Earth and Mars received foreign material throughout geologic time, when they were under heavy bombardment from rock fragments littering the solar system. During the debate over the evidence of life in ALH84001, Frank von Hippel of Columbia University and Ted von Hippel of the University of Wisconsin wrote a letter to *Science* in which they speculated about how life on Mars—if it exists or if it ever did—might have arisen. There are various possibilities, they said: Life arose on

The discovery of tube worms such as these on the cold, dark floor of the Pacific Ocean shows that life can exist in extremely inhospitable environments on Earth—suggesting that life might have existed even on Mars.

Earth or Mars and was transported to the other. Life arose somewhere else and was transported to both planets. Or life arose separately on Earth and Mars. If life originated on one of the planets and was transported to the other, it's likely life originated on Mars and was transported to Earth. "Because of the weaker gravitational field of Mars, major Martian impacts are more efficient than their Earth counterparts at pro-

pelling intact debris out of the planet's gravitational hold," they wrote. Likewise, Earth's stronger gravity makes it more efficient at sweeping up that debris.

While researchers continue to debate the questions surrounding life on Mars, it could be that one question has already been answered. If life originated on Mars and was transported to Earth, then we have already found Martian life. It is us.

Mars Global Surveyor high-resolution image of a plateau in Valles Marineris about six miles across (10 km) evidences fine layers of rocks in the walls. The layers could be volcanic rocks or sedimentary rocks laid down in water. Spectral data indicate the presence of a mineral common to volcanic rocks. Either way, the shallow crust of Mars appears much more complicated than scientists once thought.

ensured that the exploration of Mars would continue.

With Pathfinder, NASA demonstrated that it can drastically cut the cost of sending spacecraft to other planets while still generating abundant scientific findings. The price of unmanned exploration is now within a range NASA can afford. And for the first time, NASA has established a continuing program of Martian exploration. Indeed, space probes bound for Mars are rolling off an assembly line of sorts. Lockheed Martin Astronautics in Denver won the contract with NASA to build the next two Mars-bound space probes. Launches are planned every 26 months well into the next decade—there will be no more 20-year gaps between visits to Mars. In the next few years, the explorations re-invigorated by Pathfinder will continue as the agency prepares for a mission in 2005 to bring rocks back from Mars—possibly resolving, once and for all, whether the strange objects found in Martian meteorite ALH 84001 are the fossils of ancient microbes. Beyond this "sample-return" mission, as it is called, NASA has also begun preparing for the ultimate voyage to Mars—a manned mission, perhaps as early as 2015. Sagan's decades-old dream of exploring another world might finally be realized, as human footprints appear for the first time in the powdery Martian dust.

Pathfinder's immediate successor is a spacecraft known as Mars Global Surveyor, or MGS. Mars Global Surveyor was launched on November 7, 1996, 27 days before Pathfinder. Sent on a more leisurely trajectory, MGS didn't arrive at Mars until September 11, 1997, more than two months after Pathfinder. Unlike Pathfinder—an engineering mission to which scientific instruments were added later—MGS was designed from the start as a science mission. It is the most powerful spacecraft ever sent to Mars, with more computing power and more memory than JPL had ever flown before. MGS is also a kind of remedial space probe, intended to replace Mars Observer, the last of the old-style, bulky, expensive Mars missions. Mars Observer, launched in 1992, was lost as it was about to go into orbit around Mars, probably because of the rupture of a clogged fuel line. Its failure was partly what led NASA Administrator Daniel

Goldin to abandon the old-fashioned, Goliath-sized missions for a regular program of "faster, better, cheaper" Mars probes. With a regular program in place, the loss of a single mission would be damaging but not catastrophic, Goldin reasoned.

One way the costs of MGS were kept down was by using scientific instruments that had been built as spares for those carried by Mars Observer. MGS was able to fly five of the seven instruments originally carried by Mars Observer. A comparison of the cost figures for Observer and Global Surveyor provides a stark indication of the change in thinking at NASA. Mars Observer cost about $900 million; MGS cost $148 million, coming in under budget and in record time. It was built by Lockheed Martin in 26 months. The two heaviest instruments aboard Mars Observer were scheduled on later flights, to keep MGS from getting too heavy. As a result, MGS, like Pathfinder, was launched on a Delta rocket at a cost of about $50-60 million. The larger and heavier Mars Observer had been carried aloft on a Titan, at a cost of $300 million. "Global Surveyor will give us 80 percent of the Observer's science at one-quarter the cost," Goldin said when the spacecraft was launched.

As was the case with Pathfinder, however, Global Surveyor would not be entirely free of problems. After its November 1996 launch, MGS emerged from the protective launch cocoon into which it had been carefully folded and unfurled its delicate solar-panel wings. MGS's builders had put little dampers into the joints on the solar panels so they would open slowly and carefully. But one damper apparently broke and became jammed in the "shoulder" joint where one of the solar panels was attached to the rest of the spacecraft. So the solar panel didn't open to the correct locked position. The problem with the joint was a concern, but it didn't seem to pose a danger; the spacecraft seemed to be operating normally, and the flight to Mars continued on schedule.

On September 11, 1997, MGS reached Mars, fired its engines successfully, and entered orbit. The ground crew at the Jet Propulsion Laboratory celebrated. "That was where Mars Observer had been lost," recalled

Donna Shirley, the manager of NASA's Mars Exploration Program. This time, however, "the spacecraft got into orbit perfectly. It was planned for a 45-hour orbit, and it was right on 45 hours." That was not MGS's final orbit, however. Because of the severe size and weight restrictions on the spacecraft, mission engineers had devised a new way to get it into its final orbit.

As Shirley explained, in order to be sure that MGS could be flown on the less expensive Delta rocket, its designers couldn't carry too much fuel. The MGS instruments—the hand-me downs from Mars Observer—were designed to be used in a circular orbit of Mars. The orbit was designed so that as it passed over the Martian equator, the time on the surface would always be 2 p.m. "That makes the shadows right and the radiation right for all the instruments," Shirley said. But getting into that specific orbit required far more energy than did the less precise elliptical orbits of the earlier Mars orbiters, such as the Viking spacecraft. Where would engineers find that energy with their tight limits on fuel? "What Mars Global Surveyor decided to do was to go into a big orbit, 45 hours, and then drag through the upper atmosphere with each pass—something called aerobraking," Shirley said. Each pass through the orbit would slow the spacecraft ever so slightly, bringing it slightly closer to its final orbit. "In order to aerobrake effectively, they were going to use the two solar panels, and they even put flaps out on the end to give them more drag, so that they could slow down more," said Shirley. The solar panels, in addition to providing electric power, were doing double duty to slow the spacecraft, in the same way that a commercial airliner extends its flaps upon landing to slow its speed on the runway. "They needed to aerobrake fairly aggressively to synch up the orbit so it would be 2 o'clock in the afternoon," Shirley said.

But what about the jammed solar panel? Would that disrupt the plans for aerobraking? Engineers ran some tests and concluded that by turning the spacecraft around, in order not to put undue pressure on the slightly bent solar wing, they could conduct the aerobraking as planned without causing any further damage. "So they went into orbit and did their first aerobraking pass, and everything looked hunky-dory," Shirley said. "And they got deeper and deeper into the atmosphere, and all of a sudden they noticed that the panel was starting to be forced back into the right position. And everybody was very happy. And then the panel went past center. And so panic ensued—not panic, but big concern." Worried that the solar panel might snap off, the engineers quickly made a decision to fire MGS's rockets just enough to get it out of the atmosphere and stop aerobraking.

"We knew aerobraking was going to be tough, because Magellan had used it at Venus to get into its final orbit," said Shirley. "But Venus has a very stable atmosphere, and that aerobraking was done after years of observation. They really understood the spacecraft and how it worked. But to depend on it at the beginning of the mission to get you into the right orbit—when the Mars atmosphere can really move around, and you don't want to crash and burn—doing that when you haven't flown the spacecraft in that configuration is tough. And add to that this broken solar panel. It was really hairy."

MGS was supposed to arrive in its final orbit in the spring of 1998. Now, however, engineers had a problem. The aggressive aerobraking they had planned wasn't going to work. But Global Surveyor's instruments were all designed for a specific orbit; if that orbit couldn't be reached, the scientific value of the mission would be sharply curtailed. The trajectory specialists at JPL went to work, and soon they had come up with an imaginative solution: With much gentler aerobraking, they determined, they could put the spacecraft into an orbit in which it would cross the equator at 2 a.m. rather than 2 p.m. The new orbit was a mirror image of the original orbit. But the sun and radiation angles would all be the same when the spacecraft was on the daylight side of the planet, so the new orbit would be able to collect all the scientific data that was supposed to be collected during the original orbit. The gentler aerobraking would mean that the spacecraft would dip less deeply into the atmosphere with each orbit,

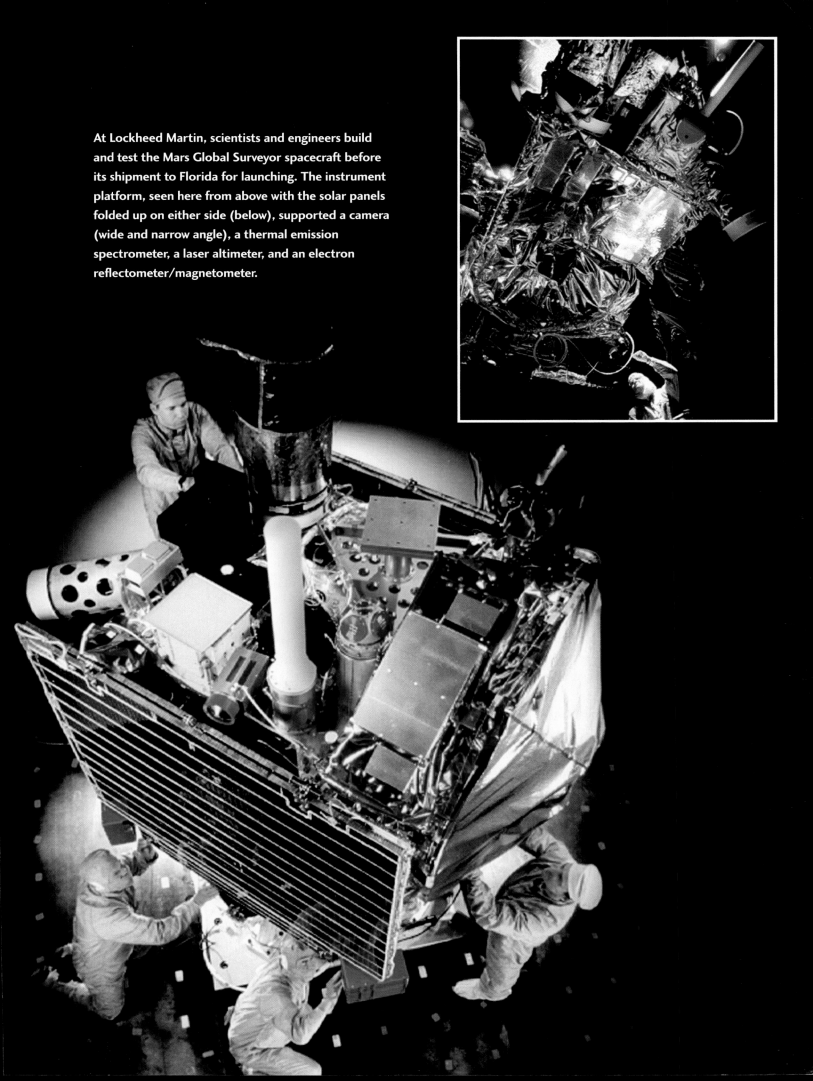

At Lockheed Martin, scientists and engineers build and test the Mars Global Surveyor spacecraft before its shipment to Florida for launching. The instrument platform, seen here from above with the solar panels folded up on either side (below), supported a camera (wide and narrow angle), a thermal emission spectrometer, a laser altimeter, and an electron reflectometer/magnetometer.

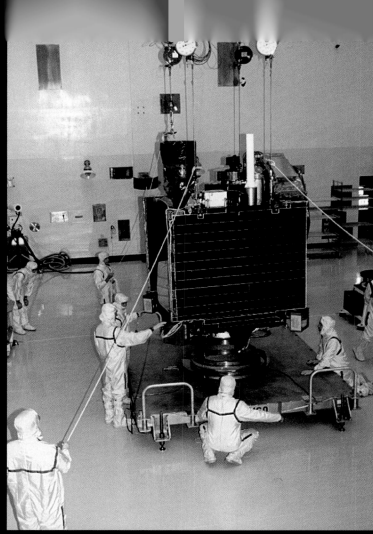

After shipment to Kennedy Space Center in Florida, the Mars Global Surveyor spacecraft undergoes processing in a clean room. Scientists and engineers install instruments onto the spacecraft (above, far left), mount the spacecraft to the launch vehicle adapter bracket (second from left), and mate the Mars Global Surveyor spacecraft to the third

and that would put less pressure on the damaged solar panel. That also meant, however, that aerobraking would take a year longer than expected. MGS would now arrive in its final orbit in March 1999.

As part of the maneuvering that would get MGS into the new orbit, the trajectory engineers determined that the spacecraft would have to stop aerobraking for three weeks in October 1997 and for about six months beginning on March 27, 1998. During those periods, the spacecraft's scientific instruments were turned on, and almost immediately scientists found the answer to one of the great unknowns about Mars: Does the planet have a magnetic field? As with so many things about Mars, this had bearing on whether or not life exists—or ever

did exist—on the red planet. While Global Surveyor was aerobraking, it was dipping closer to the planet's surface than it will be when it is in its final orbit. On those close passes—when it was about 70 miles above the Martian surface—it measured weird pockets of magnetism scattered irregularly around Mars. Michael H. Carr, the Mars expert at the U.S. Geological Survey in Menlo Park, found that to be a fascinating discovery. "The surface in places is magnetized—but only in the ancient terrain, not the modern terrain," said Carr, who headed the Viking imaging team and is a member of Global Surveyor's imaging team. "The implication is early Mars had a magnetic field and the rocks that crystallized in that field have maintained a record of that magnetic field."

stage of the Delta rocket (second from right). After the first two stages were mated on the launch pad, the third stage and the spacecraft were mounted on top. Launch (above, far right)

of the Delta II rocket from Launch Complex 17A at Cape Canaveral Air Station occurred at 12:01 p.m. EST, November 7, 1996.

The bits of magnetized rock are, in a sense, fossil evidence of an ancient magnetic field. And they were discovered only because MGS was aerobraking. "This wouldn't have been found if we hadn't had to dip into the atmosphere," Carr said. "The regular orbit is too high," and the magnetic blips are too feeble to have been detected from that distance, he said.

Before the discovery of the magnetic field, Carr said, "we assumed the interior of early Mars was hot and had a dynamo like the Earth presently does. And we assumed it cooled down early on—basically, the dynamo turned off." Earth's magnetic field is a consequence of its churning metal core, and the new finding reveals that such activity once occurred on Mars. Earlier attempts to measure a Martian mag-

netic field had shown only that if there was a magnetic field it must be very small. With the discovery that the magnetic field existed billions of years ago, researchers can begin reconstructing the "thermal history of Mars," said Arden L. Albee, chief scientist for Mars Global Surveyor. "To the extent that this all holds up, we may be able to say something about when Mars stopped being active inside, what is the age of its volcanoes—and how did this all fit together," Albee said. That is where the discovery of the magnetic field relates to the question of life on Mars. Knowing the "thermal history" of Mars could help researchers determine when liquid water might have been present and when life could have arisen.

MGS carries an extremely sensitive laser altime-

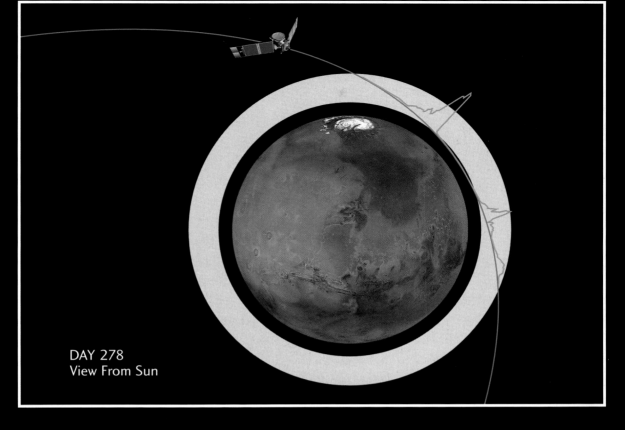

DAY 278
View From Sun

Approaching its goal, the Mars Global Surveyor aerobrakes into the planet's atmosphere in this artist's conception (above and left). The spacecraft skims into the upper atmosphere along the blue path to eventually position itself in a circular mapping orbit (dashed blue line). Aerobraking saves money by reducing the need for onboard propellant for thruster firings, allowing designers to make spacecraft small and light enough to be launched on less expensive Delta rockets. A dramatic increase in the electron concentra-tion (green line, spiking), as measured by the craft's electron reflectometer, occurred near the bottom of the ionosphere (layer of ionized gas extending from about 75 miles to several hundred miles above the surface), shown here as the shaded region, then abruptly dropped. The Surveyor also detected strong remnant magnetic fields (but no field actively generated today) below the ionosphere, locked in the crust of Mars (shown at left schematically by the magnets), suggesting an earlier convecting and molten core.

ter for measuring the height of formations on Mars to within about two yards. And the laser altimeter has shown that there are vast plains in the northern hemisphere that are exceedingly flat. They are flatter than anything on Earth—indeed, these are the flattest regions known anywhere in the solar system. "There is speculation as to why they should be so flat," said Carr. "People have suggested there may have been global oceans in these low-lying plains—and the plains could be flat because material has sedimented out of these oceans. An alternative, which I prefer, is that lakes formed at the end of the floods and the lakes simply froze in place." In Carr's view, these flat regions are the dusty surfaces of frozen lakes—the solar system's largest skating rinks.

Global Surveyor has also sent back spectacular pictures of a planet that looks far different from the one on which the Viking and Pathfinder spacecraft landed. It's almost as if MGS went into orbit around the wrong planet. Where the Viking landers and Pathfinder photographed rocky plains, Global Surveyor has found regions that look something like Utah in winter—except that they are covered with drifts of sand rather than snow. "There's a lot of evidence of dunes everywhere," said Carr. "There seems to be material blowing around everywhere. Craters are filled with dunes, and you see crescent-shaped dunes. Loose material is filling depressions and blowing around everywhere."

Global Surveyor has returned fascinating images of canyons near the Tharsis Bulge, the site of some of Mars's largest volcanoes. "Among the most spectacular pictures we've got so far are canyons," Carr said. "You can see quite distinctive layering on the canyon walls, going down kilometers. It's probably layered volcanic rocks, but it's pretty surprising....It may be that what we're seeing is that this bulge is not

FOLLOWING PAGES: **This Surveyor image of Nirgal Vallis reveals it as a valley network channel with sand dunes down the center. Sand dunes form when wind blows sand-size particles (just smaller than a millimeter across) along the surface. Flowing water probably cut the channel itself earlier in the planet's history.**

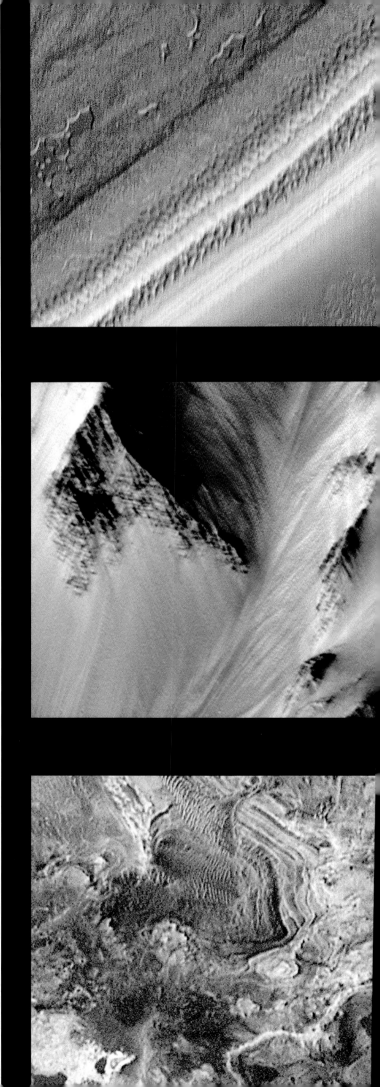

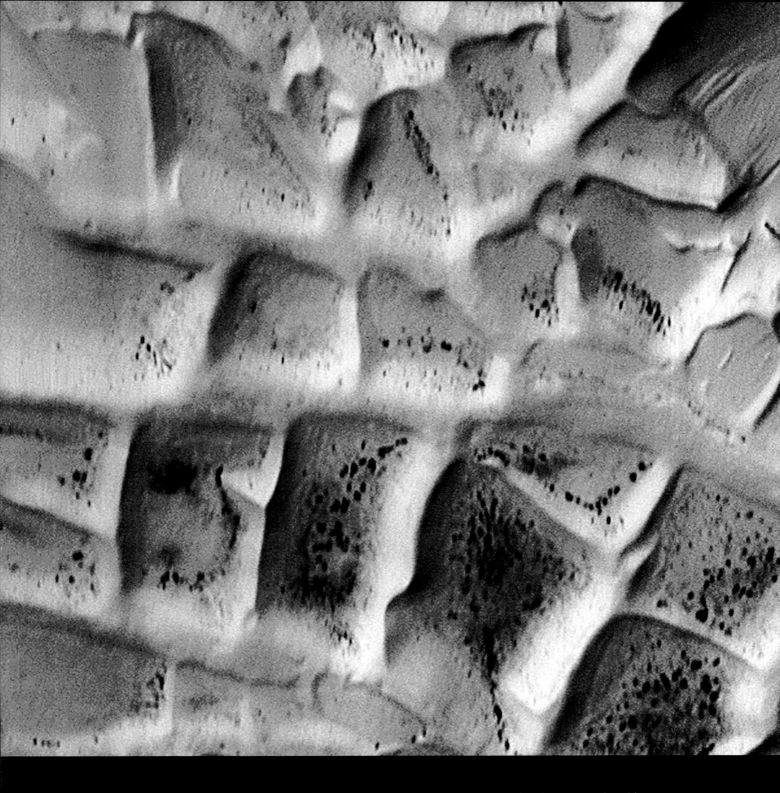

High-resolution views such as these of the Martian surface, shot by the Mars Global Surveyor camera, revealed much new information about possible forces that shaped the planet's outer layer. The finely layered material in the south polar cap (left, top) could be ice with various amounts of dust. The unusual ridges in the south polar terrain (above) hint at complex depositional and/or erosional processes. The outcrops of rock layers in western Tithonium Chasma, within Valles Marineris (left, middle) and the finely layered central deposits in the floor of Candor Chasma of Valles Marineris (left, bottom) suggest a complex history for the upper crust of Mars.

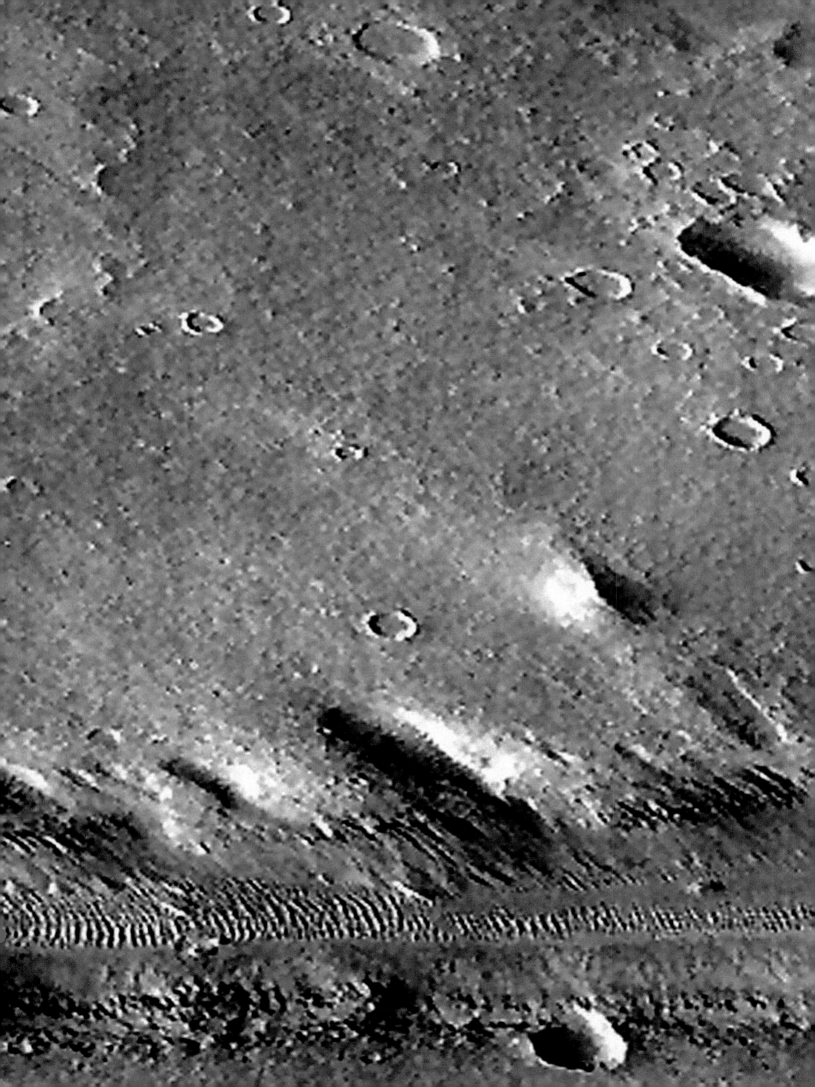

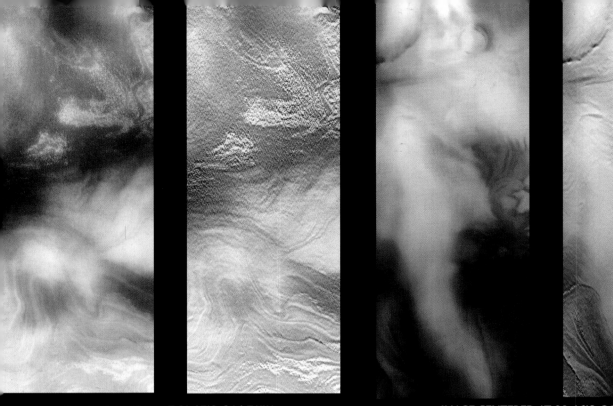

IMAGE CENTERED AT 81.97°S, 246.74°W

IMAGE CENTERED AT 80.46°S, 243.12°W

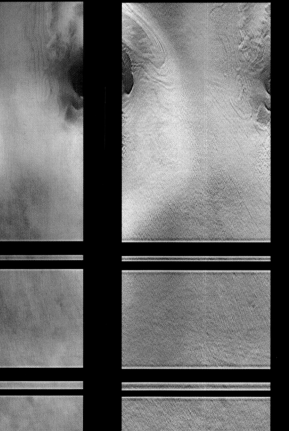

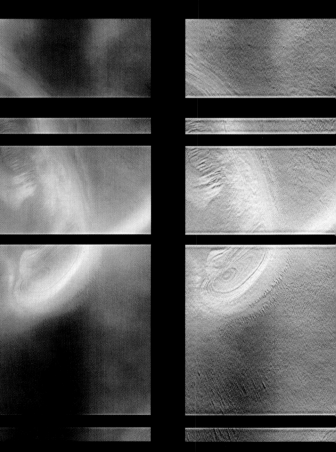

IMAGE CENTERED AT 75.8°S, 214°W.

IMAGE CENTERED AT 74.2°S, 213°W.

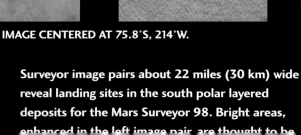

Surveyor image pairs about 22 miles (30 km) wide reveal landing sites in the south polar layered deposits for the Mars Surveyor 98. Bright areas, enhanced in the left image pair, are thought to be icy; dark areas, relatively ice free. Textures, enhanced in the right image pair, indicate ridges, pits, and troughs. The 1998 Mars Polar Lander will study fine layers for clues to climatic fluctuations

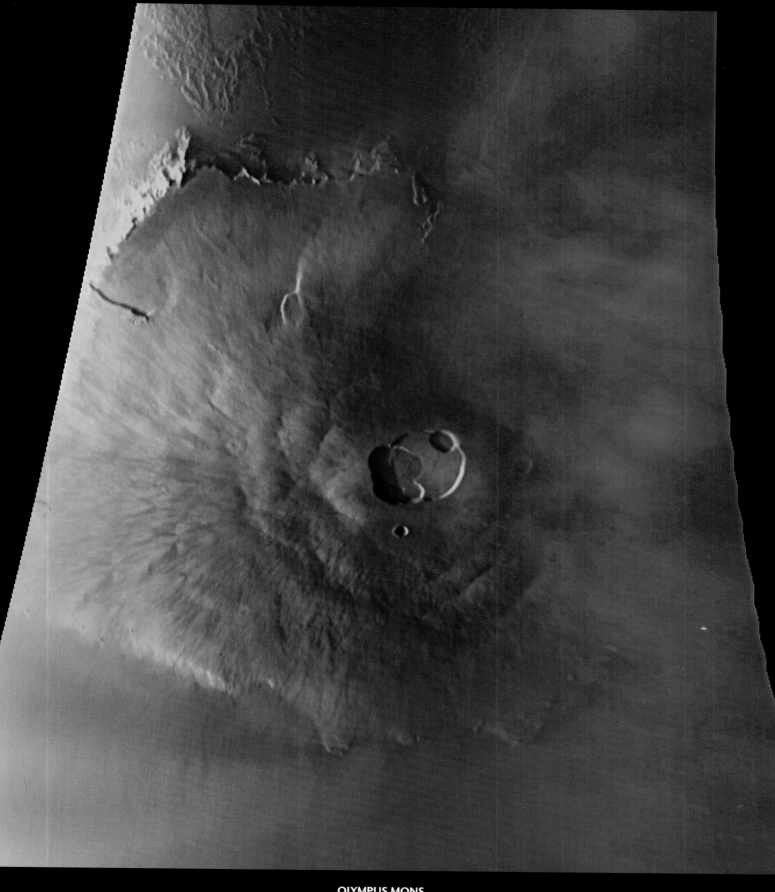

OLYMPUS MONS

Mars Global Surveyor shows in low-resolution color Olympus Mons, the largest volcano (and mountain) in the solar system. It rises nearly 17 miles (27 km) high and measures almost 400 miles (over 600 km) in diameter. Bluish water-ice clouds can be seen in the image's shadowed regions.

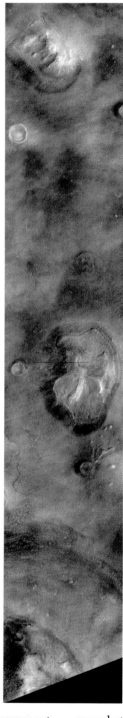

When the Mars Observer mission was lost in 1993, demonstrators gathered outside the Jet Propulsion Laboratory to expose what they believed was a government cover-up of evidence pointing to intelligent life on Mars.

The controversy involved an image from the Viking 1 orbiter taken on July 25, 1976. The spacecraft was photographing the Cydonia region of Mars, searching for possible landing sites for Viking 2. When the images were sent back to Earth and enhanced, one showed what seemed to be a mile-wide sculpture of a human face.

UFO watchers pored over the image, examining the mathematical relationships between the face and other nearby objects. Many became convinced that the Face and its surroundings were some sort of alien message. NASA offered a different theory: The "Face" was a chance arrangement of ripples and hills which, when seen in the right light, looked like a square-jawed, tight-lipped human face—the Martian counterpart of the Man in the Moon. The JPL demonstrators were convinced that better images from Mars Observer would provide evidence that the Face was created by intelligent beings. When the Mars Observer was lost, they concluded that NASA had found such evidence and was trying to suppress it. The matter rested there until April 1998, when Mars Global Surveyor began imaging Cydonia.

The Observer image is one of a series of nine showing the Face perched on an escarpment that separates cratered highlands in the south from lower, smooth plains to the north. The images are all of low or moderate resolution, and they show the Face only as a shadowy, fuzzy visage.

Mars Global Surveyor, the Observer's successor, included a much better imager capable of providing a sharper view of the Face. The spacecraft's imager was turned on briefly in the fall of 1997 and again in the spring of 1998. Shortly after midnight Pacific Standard Time on April 5, the imager obtained a high-resolution image of the Face. The raw image was transmitted to Earth later that day, processed and released on the Internet on April 6.

Mars Global Surveyor was only about 275 miles from the Face when the image was made. This time, however, the light was coming from another direction, and the result was much different. Instead of a face, the Surveyor showed an eroded knob surrounded by debris. Michael Carr, the Mars authority at the U.S. Geological Survey in Menlo Park, California, said the features are similar to others found in that part of Mars. "I don't see a face. Do you?" he asked. "I hope this [image] has scotched this thing for good."

There isn't much chance of that. Believers in the Face charged that NASA had stripped data from the new image before it was released. At least one scientist was disappointed, too. Michael Ravine, a projects manager at Malin Space Science Systems in San Diego—which built the imager for Mars Global Surveyor—said, "If we found flying saucers carved at the base of this thing...I would have been even happier. That would be cool."

The Mars Global Surveyor satellite obtained these images of the Cydonia region of Mars in April 1998 as it passed over an area that had sparked intense interest among UFO buffs.

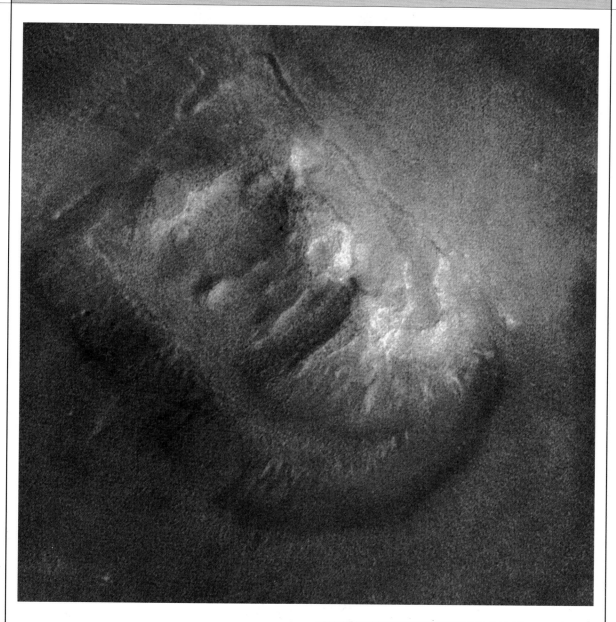

The Viking 1 orbiter captured this view of the Face on Mars (right) on July 25, 1976, as it observed Mars in search of a landing site for Viking 2. Scientists treated the image as nothing more than an aberration caused by light casting shadows across ridges on the Martian surface. Many UFO watchers believed that intelligent beings had constructed the mysterious visage, beings who either lived on Mars or had paid it a visit. A new image by the Mars Global Surveyor (above) showed that light shining obliquely on eroded ridges had produced the Face.

CONTINUING THE MARS JOURNEY

The Mars 98 mission includes an orbiter, planned to launch in December 1998 and a lander to launch a month later (above). The orbiter will study the Martian climate and the planet's surface. The lander will examine soil and ice composition in polar layered deposits at a site near the south polar cap to investigate climate change. This information will assist in planning another mission in 2005 to collect rock samples on Mars and bring them back to Earth (opposite).

something that's been pushed up, it is something that's accumulated. The layers and layers of volcanic rocks are what has caused the Tharsis Bulge."

All of this, however, is a mere preliminary to what Global Surveyor will send back when it enters its circular orbit in 1999. The spacecraft will provide the first detailed measurements of the entire planet and its atmosphere. "The data we have so far is sort of tantalizing, because it's so good," Carr said, but it's not nearly as good as what will be beamed back in 1999.

NASA now plans to take advantage of every launch opportunity to send spacecraft to Mars. The agency

will launch two more spacecraft in the winter of 1998-99, before Mars Global Surveyor reaches its final orbit. (The launch opportunities occur about every 26 months.) "For the first time in a long time, NASA has a program of missions with a unified objective," said Mark Adler, a Jet Propulsion Laboratory engineer with the title of Mars Exploration program architect. As he explained, that puts him in charge of designing the program. The Mars Surveyor Program, as the exploration program is called, was initially given a budget of 200 million dollars a year. That rose to 250 million dollars when NASA scientists at the Johnson

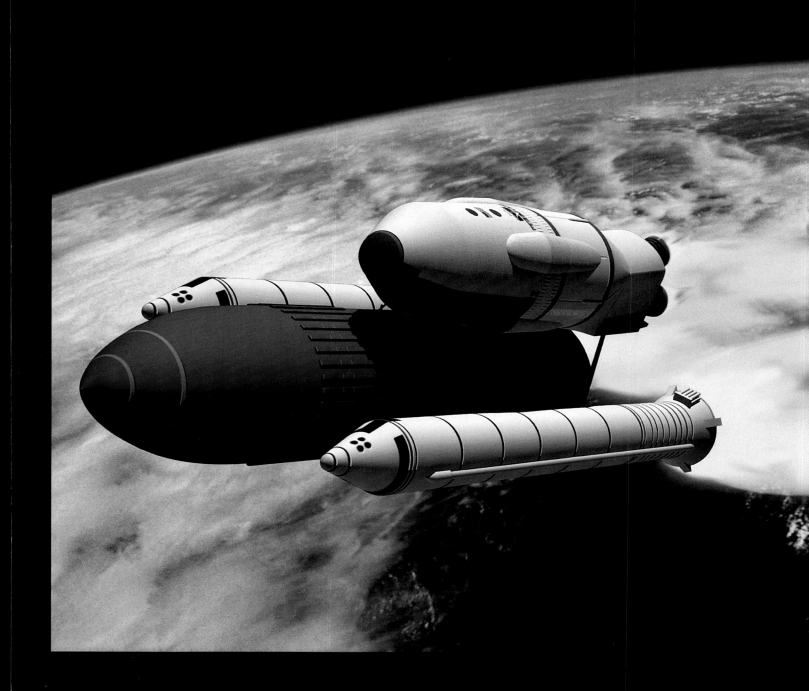

Solid rockets, based upon those used for the space shuttle, might be a crucial component of a mission to send a human crew to Mars. As bringing the costs down to a manageable figure could help make such a voyage possible, the use of existing space hardware might factor into the equation. This artist's conception shows one of the two launch vehicles that would be needed to put together a mission to send astronauts to Mars.

Space Center announced, in August 1996, the possible discovery of life in the Martian meteorite ALH 84001. "After ALH 84001, we were asked specifically how we would change the program to search for life," Adler said. "So over a couple of days, we came up with an answer. We provided a bunch of alternatives with different funding levels." NASA administrators decided on a 50 million dollar annual increase to expand the search for life on Mars. "The rock had a very significant impact on our program, in terms of augmented funding to pursue more ambitious goals."

In December 1998, NASA will launch the Mars Climate Orbiter, a spacecraft designed to study daily weather and atmospheric changes, monitor water vapor in the atmosphere, and look for evidence of past climate change, among other things. The Climate Orbiter is paired with the Mars Polar Lander, which will be launched in January 1999. The orbiter carries an atmospheric monitoring instrument called the Pressure Modulator Infrared Radiometer, or PMIRR. This is a duplicate of one of the instruments from the failed Mars Observer mission that was too heavy to be carried aboard Mars Global Surveyor. The primary job of the PMIRR is to look for water in the Mars atmosphere, Adler explained. "Water is the unifying theme throughout the exploration of Mars," he said. "The whole water theme is tied up in understanding the climate, understanding how water moves around Mars, understanding the resources available on Mars and where is the water, and understanding the origin of life. Because where there was water, there may have been life that originated on Mars or was transported to Mars."

The 1998 and 1999 missions were devised before the Martian meteorite announcement, but even then their goal was to study water on Mars. "The orbiter is going to look at water in the atmosphere, and the 1999 lander mission will be going near one of the poles of Mars, where there is frozen water on the surface—or within a few centimeters of the surface," Adler said. "Water cannot exist near the equator of Mars because it's too warm, and water sublimates out of the soil." That is, like dry ice on Earth, water ice becomes a gas when it melts, not a liquid.

The 1999 lander will look more like Viking than Pathfinder. It will land on legs, using rockets, not by means of the bouncing balloon package that got Pathfinder safely to the ground. (The air bag landing system might one day be used again, though. "It's a resource, something that's out there that we know how to do, and we can do it if we want to," said Adler.) Nevertheless, much of the engineering on the 1999 lander was derived from Pathfinder. The heat shield, radar, and parachute all came from Pathfinder, Adler said. "We could not have done this mission as cheaply as we're doing it if it hadn't been for Pathfinder. The only thing that's different is we've replaced the air bags with rockets and legs. We're using the Pathfinder entry and descent, but not landing." Adler said.

The 1999 lander will also carry along a piggyback mission called Deep Space-2. That mission consists of two novel devices known as "microprobes," which are designed to bore into the Martian surface. While the lander is still high above the Martian surface, it will release two basketball-sized objects that will then plummet to the planet's surface. When each hits the ground, the outer aeroshell that protected it during descent through the Martian atmosphere will shatter, allowing a device inside called a microprobe to burrow into the Martian surface. Each bullet-shaped microprobe is about 4 inches long and designed to penetrate as far as two yards into the surface. It is attached by a cable to a slightly smaller, disk-shaped transmitting station which will remain on the surface. Instruments in the microprobes will gather soil and measure temperature, water content, and other properties. The data will be transmitted from the surface to Mars Global Surveyor, and from there to Earth. The primary objective of the Deep Space-2 microprobes is to determine how much water might be concealed as ice beneath the Martian surface. If the microprobes prove effective, they will be incorporated into future planetary missions, said Deep Space-2's project manager, Sarah Gavit of JPL.

The next pair of low-budget spacecraft in the Mars exploration program will be launched at the next opportunity—in 2001. Those spacecraft will begin collecting data for the next large-scale Mars mission—the attempt in 2005 to collect rocks on the Martian surface and bring them back to Earth for study. "We're focusing a fair bit of our resources on getting science return out of rocks we bring back," Adler said. The orbiter will measure radiation on Mars and investigate the mineral composition of the Martian surface. The 2001 lander is designed to study soil, atmospheric chemistry, and radiation at the surface. One of the goals of the two spacecraft, Adler said, "is to help us select good places to put rovers down so we can find the kinds of rocks we're looking for." An instrument called the Thermal Emission Imaging System, or THEMIS, "should be able to tell us the kind of rocks you'll find on these various places on the surface," Adler said. "Based on the geology and a lot of inference, we can say that we think an area will have a certain kind of rocks that have undergone rapid mineralization"—that is, rocks likely to contain fossils. "Then you have to go down and find those rocks. That's what rovers are for."

Another orbiter-lander pair is scheduled for launch in 2003, to search for rocks in yet another site. The 2003 lander will include a rover to search for rocks. Part of their job is to look for three kinds of sites where life, or evidence of past life, might be found. These include places where water might remain liquid beneath the surface, places where liquid water once existed on the surface, and sites where water and the heat of volcanic activity might have once created a spawning ground for life. The last, the so-called ancient hydrothermal environments, are likely now buried beneath the Martian surface, Adler said. "So we look for places where very large meteors have hit Mars and dug up stuff that's kilometers deep and thrown it out. We find these craters with these ejecta blankets and say, ah, these ejecta blankets may have what we're looking for."

In 2005, then, NASA will launch its first sample-return mission, an ambitious project to bring Martian rock samples back to Earth. "It's almost immeasurable how much more you can do with rocks that are here compared to rocks that are there,"

Adler said. "When we send stuff to Mars, we can send only these little instrument packages…. In labs on Earth, you can have rooms full of equipment that is very complex, and you're able to do far more than you could with what you can send to Mars." Even more important, when the inevitable questions are raised about the findings in the samples, those questions can be addressed immediately on Earth. When the questions are raised about findings from Mars, it sometimes requires another mission to answer them. As Adler points out, "That could take four years and cost a fair chunk of change."

The mission will bring back about one pound of Martian rock samples. "But what's important is that it will consist of 30 to 100 different kinds of rock," he said. "It's the diversity that gives you any chance of finding what you're looking for. If we brought back one half-pound rock, we might not get anything at all." On Earth, laboratory analysis can be done on minute chips of rock weighing millionths of a gram. "So we can reserve large portions of one 3-gram piece of rock [about the mass of a U.S. penny] for future study and still have enough for hundreds or thousands of investigations of that rock," Adler said.

NASA asked the National Academy of Sciences to review its proposal for a sample-return mission, and specifically to consider whether samples brought back from Mars could pose any hazard to the Earth. The academy concluded that contamination of the Earth by Martian microbes "is unlikely to pose a risk of significant ecological impact or other significant harmful effects," But, it added, "the risk is not zero." Any Martian organisms that might escape on Earth would be unlikely to overwhelm terrestrial microbes, which would be better adapted to life on the Earth. Nevertheless, the academy concluded, each sample brought back from Mars should be treated as though it contained potentially harmful organisms, until it can be proved safe.

If the sample-return mission is successful, additional sample-return missions could search for rocks in different environments than the one chosen for the first mission. Will the missions bring back the

nternational space station will be used to
 the consequences on astronauts of extended
 in space. This information will help deter-
 what astronauts might experience during an
ded stay on Mars. The largest international
ific venture ever undertaken, the space
n will be positioned at an altitude of about
iles in an orbit that will allow it to be easily
ed by the launch vehicles of all of the
ational partners. The completed space

station will be 356 feet across and 290 feet long
and will accommodate up to six people at a time.
The podlike structure suspended from the station
at the bottom of this illustration of the completed
station is a Mars habitat module, which could be
tested in orbit before being shipped to Mars. The
module, where astronauts could live during their
stay on Mars, is one of a variety of components of
a human Mars mission that could be tested on the
space station.

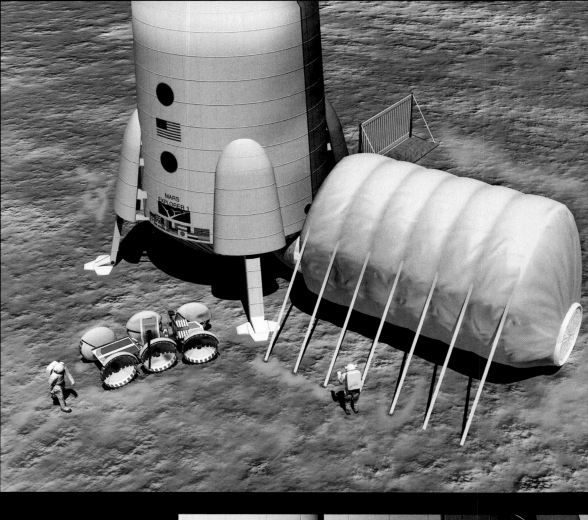

On Mars, a human crew could live in a module such as this (above). To reduce the costs of getting the crew to Mars and safely back again, the base station on Mars could be equipped with machinery for making rocket fuel to power the return trip to Earth. An inflatable laboratory (right) might be included to expand the base station, giving astronauts more room to work and live in a pressurized environment.

gold—will they return evidence of life on Mars? "If there was never any life on Mars, they're not going to bring anything on back," Adler said. "If Mars was absolutely teeming with life at one point, just covered with life, then the chances are very good we'll bring something back. But if life was rare on Mars, which is probably the most likely thing, we probably don't have a very good chance of bringing anything back. But in the process of looking, we're going to learn a lot.... I think we'll get a tremendous increase in our understanding of Mars, while at the same time maximizing our chance—whatever that chance is, and it may be small—of bringing back the right kind of rocks if they are there for us to find."

If interest in the exploration of Mars remains high during the early years of the 21st century, the completion of the sample-return missions will leave one mission as the next logical step in the exploration of Mars—a mission to send astronauts to Mars. The planning has already begun. "We're trying to figure out how to explore Mars with robots and humans and how to lead towards human exploration of Mars around the 2014-or-so time frame," Adler said. The question is: Can the human exploration of Mars be done at a cost that Americans are willing to pay?

One way to keep costs down is to keep the weight of the manned spacecraft as low as possible, Adler said. "That's true for all missions," he said. "For human missions to Mars, one way to do that is to produce the propellant that you need on Mars instead of bringing it with you. So you can send equipment to suck in Mars atmosphere, which is carbon dioxide, and use it to produce oxygen. You can use that to fuel up your spacecraft for launch into Mars orbit. But of course, that's very risky. If your return trip depends on this equipment working, you want to make sure it works."

A demonstration of the technology is already being planned. NASA wants to send a small fuel production system to Mars to see whether the idea works. "It will actually be taking in carbon dioxide and trying to produce small amounts of oxygen at a very small scale, nowhere near the scale you'd need to return people," Adler said. "But it would be some-

thing that shows that the idea can work." Included with that will be instruments to analyze Martian dust to see whether it can be damaging to equipment or to people. "You want to understand what the characteristics of the dirt are," Adler said. "Is it going to gum up your equipment? Is the stuff toxic? How do you have to design your equipment to survive in the environment that has a lot of dust? Mars has dust storms, dust is everywhere."

A radiation experiment is also planned to determine exactly how much and what kind of radiation astronauts on Mars might be exposed to. "That will tell us how much you have to protect humans on Mars, how long they can live there and what their added cancer rate is going to be because they live there," Adler said. Scheduling of a human mission is highly tentative; NASA changes its plans often, and its long-range forecasts are notorious for being inspiring, imaginative—and rarely followed. But Adler envisions a possible cargo mission to Mars in 2011 that would deliver the equipment needed to produce fuel from carbon dioxide. The human mission would be launched three years later—in 2014. The idea of the tentative scheduling is to help determine what kinds of technology must be developed for the mission to be a success. "We're laying the groundwork now," Adler said.

In 1989, President George Bush went to the National Air and Space Museum on the 20th anniversary of the first manned moon landing to call for a reinvigoration of the U.S. space program—including sending astronauts to Mars. No one knew what such a mission would cost. Some guessed it might be as much as 800 billion dollars. A government report was commissioned to try quickly to assess the cost. This so-called "90-day Report" put the cost at 450 billion. At the time, Robert Zubrin was a senior engineer for Martin Marietta Astronautics Company, which is now part of Lockheed Martin. To Zubrin, that figure sounded far off the mark. He decided to devise his own estimate of the cost of a mission to Mars. That marked the birth of a program he calls "Mars Direct," a new approach to the exploration of Mars. Zubrin detailed the plan in his 1996 book, *The Case for Mars:*

The Plan to Settle the Red Planet and Why We Must.

"This plan employs no immense interplanetary spaceships, and thus requires neither orbiting space bases nor storage facilities," Zubrin explained. "Instead, a crew and their habitat are sent directly to Mars by the upper stage of the same booster rocket that lifts them to Earth orbit, in just the same way as the Apollo missions and all unmanned interplanetary probes launched to date have flown." The key to the plan, Zubrin noted, is "the mission's ability to use Mars-native resources" to make its return fuel on the surface of Mars. "It is this concept of 'living off the land' that makes Mars Direct possible," he wrote. "Consider what would have happened if Lewis and Clark had decided to bring all the food, water, and fodder needed for their transcontinental journey. Hundreds of wagons would have been required....A logistics nightmare would have been created that would have sent the costs of the expedition beyond the resources of the America of Jefferson's time."

Zubrin first put forward his ideas in 1990, and since then he has campaigned vigorously for their adoption. The Mars Direct mission would put humans on Mars as early as 2008. And the cost? No more than 30 billion dollars, and perhaps closer to 20 billion—a tiny fraction of the 450 billion estimate made during the Bush administration. "Twenty to thirty billion dollars is not cheap," Zubrin wrote, "but it's roughly in the same range as a single major military procurement for a new weapons system." If it were spread over 20 years—with the first 10 years for building equipment and the second decade for flying missions—the costs would represent only 8-12 percent of NASA's budget, he said. "For the sake of opening a new world to human civilization, it's a sum that this country can easily afford," he wrote.

Here is how an accelerated version of the mission might look, putting astronauts on Mars in 2008: The workhorse launch vehicle would be a rocket using space shuttle engines and their solid rocket boosters. In 2005, one of these rockets would send a 50-ton payload to Mars: the Earth-return vehicle. It would arrive with a nuclear power supply and empty fuel tanks. As soon as it landed, it would begin making fuel. Rovers would search the area for the best landing site, and when they found it, they would plant a radar beacon. In 2007, a second Earth-return vehicle would be launched. A few weeks later, four humans would soar aloft on the first journey to Mars. They would be in a habitation module with three years' worth of food and a life-support system able to recycle oxygen and water. Also on board would be a car for exploring Mars.

After a six-month flight, the astronauts would land on Mars in 2008. The second Earth-return vehicle, sent on a slower trajectory, would arrive soon afterward to provide a backup—and to carry the next crew home if it were not needed by this one. The astronauts would spend 500 days on Mars, searching for life, studying Martian geology and prospecting for minerals. Their car would use some of the fuel that had been manufactured by the first Earth-return vehicle. When they began their six-month return trip to Earth, they would leave behind the habitation module and the second Earth-return vehicle for the next crew.

The mission has risks. The consequences of extended exposure to the low gravity on Mars, to radiation, and to the Martian environment are unknown, Zubrin noted. But the risks will remain "whether we make the attempt with Mars Direct in 2007 or leave it for another generation to try," he said.

Not everyone shares Zubrin's vision. Donna Shirley, the head of NASA's Mars exploration program, worried that Zubrin makes it all sound too simple. "He wishes away all the technical problems—it ain't that easy," she said. Others believe the costs could be far higher than Zubrin's estimates. Whether the particulars of his plan are accurate or not, Zubrin has helped open the door to a new way of thinking about the human exploration of Mars.

Mars Direct is a plan to explore Mars over the next couple of decades or so. But what happens then? The surface of Mars is about the same size as the continents on Earth, and, as Carl Sagan has pointed out, exploring the planet could take centuries. But at some time in the distant future, the job may be completed.

"If there is life on Mars, I believe we should do nothing with Mars," Sagan wrote in *Cosmos*. "Mars then belongs to the Martians, even if the Martians are only microbes." He then asked what might happen if Mars were found to be lifeless: Could we transform Mars into a habitable planet?

The idea is called terraforming—the conversion of Mars into an Earthlike place. Sagan is not the only scientist to give thought to the idea. NASA has itself supported some research on terraforming Mars. If Mars were to become more Earthlike, it would need a thicker atmosphere, more oxygen, and liquid water. A thicker atmosphere would prevent much of the Sun's harmful ultraviolet radiation from reaching the surface, just as Earth's atmosphere does. The polar caps might be melted, although that would require enormous amounts of energy. The melting of the polar caps could aid the formation of an atmosphere. Genetically engineered plants might be used to help transform the planet.

The process would likely begin with efforts to raise temperatures. Zubrin listed three ways that might be done. One would be to put mirrors into orbit around Mars, pointing them so they would reflect sunlight onto the planet's surface. Another idea is to build factories on Mars to produce greenhouse gases. Among the most potent greenhouse gases are the chlorofluorocarbons that have until now been used in refrigeration and air-conditioning equipment. They are being phased out on Earth because of their harmful consequences for global warming and the depletion of the ozone layer. But the same thing that makes them dangerous on Earth could make them useful in a program to terraform Mars. Producing CFCs on Mars would require a huge industrial facility. Zubrin estimates it would take several thousand workers and need about as much power as is used today by the city of Chicago. "Nevertheless, all things considered, such an operation is hardly likely to be beyond the capabilities of the mid-twenty-first century," he concluded.

A third idea, Zubrin said, would be to use biological organisms to terraform Mars. Bacteria could be used on Mars to produce ammonia and methane, both powerful greenhouse gases. The planet would need to be warmed first by orbiting mirrors or CFCs, but once conditions on the planet had changed sufficiently to allow liquid water to exist on the surface, microbes could be brought in to complete the job.

While the challenges and costs would be staggering, the calculations show, Zubrin said, that the technology exists to terraform Mars. "Only human explorers operating on Mars can learn enough about the planet and the methods of utilizing its resources to turn such a dream into reality," he said. But the effort is worth making, he said, because "what is at stake is an entire world."

Sagan has concluded that terraforming Mars could take hundreds or thousands of years. And it would probably require a variety of other technologies in addition to those used to warm the planet. "We might wish not only to increase the total atmospheric pressure and make liquid water possible but also to carry liquid water from the melting polar caps to the warmer equatorial regions," Sagan wrote. "There is, of course, a way to do it: we could build canals."

In that unimaginably distant future, then, Percival Lowell's notion of canals on Mars might finally come true. The canals that Lowell imagined might one day actually exist on the Martian surface. And, of course, Lowell would be proved correct about the existence of life on Mars, too. The terraforming of Mars would likely be done by people whose permanent residence and primary affiliation is with Mars, Sagan noted. "The Martians," he said, "will be us." Or, as T. S. Eliot wrote:

We shall not cease from exploration
And the end of all our exploring
Will be to arrive where we started
And know the place for the first time.

FOLLOWING PAGES: **This well-known painting by Pat Rawlings illustrates the possibility of humans someday—perhaps in the not too distant future—walking on the surface of Mars, a planet in some ways similar to our own Earth.**

Mars & the Solar System

The solar system was formed about 4.5 billion years ago, as grains of dust coalesced into bigger and bigger objects until they formed the planets. The terrestrial planets—Mercury, Venus, Mars and the Earth—differ markedly from the gaseous giants Jupiter, Saturn, Uranus and Neptune. And they all differ from Pluto, the farthest planet from the Sun and the smallest of them all. In addition to the nine planets, the solar system includes dozens of moons and countless asteroids, comets and meteorites. Here is a quick tour of the other bodies in the solar system aside from Mars and the Earth:

● The Sun's diameter is about 100 times that of the Earth, and although it is far less dense than the Earth, its mass is more than 300,000 times greater. The temperature at the center of the Sun is about 27 million degrees.

● Mercury is the second-smallest planet, after Pluto. It revolves around the Sun every 88 days. It has some of the most extreme temperatures in the solar system, with a maximum surface temperature of 800°F and a minimum temperature of minus 280°F.

● Venus, like Mercury, has no moon. It is nearly the same size as the Earth, and although it is nearly twice as far from the Sun as Mercury, its dense atmosphere captures so much of the Sun's heat that the mean surface temperature is about 850°F.

● Jupiter is the largest of the planets, with a diameter about 11 times that of the Earth, but it is still far smaller than the Sun. It has a ring system, four principal moons—Io, Europa, Ganymede and Callisto, and a dozen smaller ones.

● Saturn, which is slightly smaller than Jupiter, has 18 moons and the most extensive ring system of any of the planets. One of its rings, the F ring, is made of two narrow bright rings with a fainter ring inside them, all of which combine to give it a braided appearance.

● Uranus is another giant ringed planet with more than a dozen moons, though it is less than half the size of Jupiter and Saturn. Its atmosphere has a greenish color produced by methane and high-altitude smog. One of the rings of Uranus is believed to be kept in place by the gravity of two "shepherd" moons on either side of it.

● Neptune, which is a near twin of Uranus in size and density, is about one-and-a-half times as far from the Sun as Uranus. Its atmosphere includes distinctive moving features with names like the Great Dark Spot, Dark Spot 2 and the fast-moving Scooter. Winds can reach speeds of 400 miles per hour.

● Pluto's diameter is less than one-fifth that of the Earth, and it has a moon, Charon, that is about half its size. The Hubble telescope has shown that Charon is bluer than Pluto, meaning the two objects have different compositions.

● The solar system is populated with untold numbers of comets, asteroids, and meteorites. The comet Hale-Bopp was one of the most impressive smaller objects to pass near Earth in recent years, and Halley's the most famous. One 32-mile-long asteroid, called Ida, was found recently to have its own tiny moon, named Dactyl. The Galileo spacecraft obtained images of the pair as it passed by them on August 28, 1993.

● Mars is about half the size of the Earth, but it is less dense and weighs only about one-tenth as much.

● The Martian surface covers about as much area as the continents on Earth.

● Astronauts on Mars would weigh about one-third what they weigh on Earth.

● The Martian atmosphere is about 95 percent carbon dioxide and 3 percent nitrogen. Earth's atmosphere is 78 percent nitrogen and 21 percent oxygen.

● The daytime high temperature at the surface of Mars is 65°F, but the temperature drops dramatically only a few feet above the surface. The daytime high 5 feet above the surface is only 15°F. At night, temperatures drop to a low of minus 130°F at the surface and minus 105°F at five feet.

● The Martian axis is tilted at about the same angle as Earth's axis, so Mars has seasons similar to those on Earth. Its elliptical orbit, in contrast to Earth's nearly circular orbit, means its seasons are also affected by its distance from the Sun.

FINDING MARS IN THE NIGHT SKY

As the Greeks observed, the planets wander among the fixed stars in patterns that change over time, so Mars isn't always visible. It can be viewed all night when it is in opposition, meaning that it is on the opposite side of the Earth from the Sun. During opposition, Mars is highest in the sky at midnight. The dark areas on Mars and the polar caps are not visible without a medium-power telescope.

The position of Mars with respect to the constellations changes, but Mars always lies somewhere in the constellations that make up the zodiac. It is one of the brightest objects in the sky, and it is at its brightest near opposition, when the Earth and Mars are closest together. A current star map is needed to get the exact location of Mars, but two characteristics make Mars relatively easy to spot: It has a distinct reddish hue when seen with the naked eye. And because it is a planet, it does not twinkle. That helps distinguish it from red stars such as Aldebaran (in the constellation Taurus) and Antares (in the constellation Scorpius).

A Comparison of Mars and the Earth

	EARTH	MARS
DIAMETER	7,926 miles	4,219 miles
LENGTH OF YEAR	365 days	687 days
LENGTH OF DAY	24 hours	24.6 hours
DISTANCE FROM SUN	91-94 million miles	128-155 million miles
ATMOSPHERIC PRESSURE	1014 millibars	7-9 millibars
ESCAPE VELOCITY	25,055 mph	11,185 mph

(the minimum velocity to escape the planet's gravitational pull)

INDEX (Bold numerals indicate illustrations)

LIBRARY OF CONGRESS CATALOGING-IN-PUBLICATION DATA

Raeburn, Paul.
    Uncovering the secrets of the red planet : Mars / by Paul Raeburn
; foreword and commentary by Matt Golombek.
        p.   cm.
    Includes index.
    ISBN 0-7922-7373-7 (reg.) ISBN 0-7922-7031-2 (dlx.)
    1. Mars (Planet) 2. Mars (Planet)—Exploration. I. Title.
QB641.R34 1998
523.43—dc21

           98-13991
           CIP

ACKNOWLEDGMENTS

One evening while I was working on MARS, I took a break and went to the local bookstore. As I browsed, looking forward to the day when I would be finished and this book would join the others on the shelves, I began to think about how much work went into making the products that were for sale in that single large bookstore. Very few things are as hard to make as a book, and there before me were as many as 100,000 different books—each of them the product of months or years of effort. Collectively, they represented millions and millions of hours of work.

That made me think about the hours I was spending on MARS, and how they were more productive and pleasant than they might otherwise have been, thanks to the help and cooperation of so many talented people. The scientists and public relations professionals at the Jet Propulsion Laboratory were unfailingly generous with their time. I would like especially to thank Matthew Golombek for talking to me about Martian science, for contributing the foreword and most of the captions in this book, and for providing valuable feedback on the manuscript. Among the many other Pathfinder scientists and engineers who took time to talk to me and to read portions of the manuscript are Richard Cook, Rob Manning, Jennifer Harris, Tim Parker, and Mark Adler.

My Editor at the National Geographic Society, Kevin Mulroy, and Assistant Editor, Kevin Craig, were good companions on this trip. I'd also like to thank Charles Kogod for wandering through history and the solar system to find the spectacular images in this book, and George Bounelis and Sean Mohn, for cartographic assistance.

I could not have undertaken this project without the support of my editors and colleagues at *Business Week,* especially Editor-in-Chief Stephen B. Shepard and Assistant Managing Editor G. David Wallace.

My wife, Liz, and my children, Matt, Alex, and Alicia, endured my long and frequent absences and shared my excitement about the book, for which I thank them.

And I would like to thank my parents, who remain my most uncritical and enthusiastic fans.

ADDITIONAL READING

Many fine books have been written about Mars since the time of Percival Lowell. Among those that proved indispensable in the writing of this book were four comparatively recent volumes: *Destination Mars: In Art, Myth and Science* by Jay Barbree and Martin Caidin with Susan Wright; *Journey into Space: The First Thirty Years of Space Exploration* by Bruce Murray; William Sheehan's *The Planet Mars, A History of Observation and Discovery;* and John Noble Wilford's *Mars Beckons: The Mysteries, the Challenges, the Expectations of Our Next Great Adventure in Space.*

Also useful were Victor R. Baker, *The Channels of Mars;* Ray Bradbury, Arthur C. Clarke, Bruce Murray, Carl Sagan, Walter Sullivan, *Mars and the Mind of Man; The Search for Life on Mars: Evolution of an Idea* by Henry S. F. Cooper; *To Utopia and Back: The Search for Life in the Solar System* by Norman H. Horowitz; William Graves Hoyt, *Lowell and Mars;* Clayton R. Koppes, *JPL and* the American Space Program: A History of the Jet Propulsion Laboratory; David Morrison, Exploring Planetary Worlds; The Cambridge Illustrated History of Astronomy edited by Michael Hoskin; and The Near Planets, by the editors of Time-Life Books.

The writings of the late Carl Sagan on Mars and many other scientific subjects are without equal. He was the author, co-author, or editor of more than 20 books, and his 1977 book, The Dragons of Eden, won a Pulitzer Prize. The chapter on Mars in Cosmos, written after the Viking mission, is especially recommended. Other fine examples of his work include Broca's Brain: Reflections on the Romance of Science; and Pale Blue Dot: A Vision of the Human Future in Space. A discussion of Sagan's achievements in science, education, and public policy appears in Carl Sagan's Universe, edited by Yervant Terzian and Elizabeth Bilson.

Scores of science fiction books have been written about Mars. Among the best are Ray Bradbury's The Martian Chronicles; the Martian tales of Edgar Rice Burroughs, beginning with A Princess of Mars; Kim Stanley Robinson's trilogy, Red Mars, Green Mars, and Blue Mars; and the H.G. Wells classic, The War of the Worlds.

Additional books of interest include Eric Burgess, To the Red Planet; Michael H. Carr, The Surface of Mars and Water on Mars; Maurice A. Finocchiaro, Galileo on the World Systems; Roy A. Gallant, National Geographic Picture Atlas of Our Universe; Donald Goldsmith, The Hunt for Life on Mars; H. H. Kieffer, B. M. Jakosky, C. W. Snyder, and M. S. Matthews, Mars; Percival Lowell, Mars and its Canals; Patrick Moore, Guide to Mars; and Robert Zubrin with Richard Wagner, The Case for Mars: The Plan to Settle the Red Planet and Why We Must.

The National Aeronautics and Space Administration (NASA), an agency of the United States government, has produced many worthwhile publications about space research and exploration. Periodicals that cover Mars, the solar system, space exploration and technology, exobiology, and other topics touched on in this book include Air & Space Smithsonian, Astronomy, Aviation Week & Space Technology, NATIONAL GEOGRAPHIC, Nature, Science, Science News, Scientific American, Sky and Telescope.

The World Wide Web is an invaluable source of information about Mars. It includes a vast storehouse of images from Mars missions, beginning with Mariner 4 and continuing to Mars Global Surveyor. Many spacecraft images are now routinely made available on the Web as soon as they are received and processed, so the public can see them as soon as NASA and JPL scientists do. Here are some of the best offerings on the Web:

JPL MARS PATHFINDER HOME PAGE
http://mpfwww.jpl.nasa.gov/index1.html

MARS GLOBAL SURVEYOR HOME PAGE
http://nssdc.gsfc.nasa.gov/planetary/marsurv.html

JET PROPULSION LABORATORY HOME PAGE
http://www.jpl.nasa.gov/

JPL MARS MISSIONS HOME PAGE
http://www.jpl.nasa.gov/marsnews/

NASA HOME PAGE
http://www.nasa.gov/

NASA CENTER FOR MARS EXPLORATION
http://cmex-www.arc.nasa.gov/

NATIONAL SPACE SCIENCE DATA CENTER PLANETARY HOME PAGE
http://nssdc.gsfc.nasa.gov/planetary/planetary_home.html

NASA OFFICE OF SPACE SCIENCE HOME PAGE
http://www.hq.nasa.gov/office/oss/

JOHNSON SPACE CENTER MARS METEORITE HOME PAGE
http://cass.jsc.nasa.gov/lpi/meteorites/mars_meteorite.html

LUNAR AND PLANETARY INSTITUTE MARS PAGE
http://cass.jsc.nasa.gov/expmars/expmars.html

MALIN SPACE SCIENCE SYSTEMS HOME PAGE
http://www.msss.com/

FEDERATION OF AMERICAN SCIENTISTS' MARS PAGES
http://www.fas.org/mars/index.html
http://www.fas.org/mars/missions.htm

THE GALILEO PROJECT
http://es.rice.edu/ES/humsoc/Galileo/

THE CASE FOR MARS HOME PAGE ON THE EXPLORATION AND COLONIZATION OF MARS
http://spot.colorado.edu/marscase/home.html

THE WHOLE MARS CATALOG
http://www.reston.com/astro/mars/catalog.html

RUSSIAN SPACE RESEARCH INSTITUTE HOME PAGE
http://www.iki.rssi.ru/

COVER IMAGE FROM MARS: UNCOVERING THE SECRETS OF THE RED PLANET
http://www.atmos.washington.edu/mars/special/334sp/me07s078.gif

ILLUSTRATION CREDITS

MARS
UNCOVERING THE SECRETS
OF THE RED PLANET

Paul Raeburn

Foreword and Commentary by Matt Golombek

PUBLISHED BY THE NATIONAL GEOGRAPHIC SOCIETY

John M. Fahey, Jr., PRESIDENT AND CHIEF EXECUTIVE OFFICER

Gilbert M. Grosvenor, CHAIRMAN OF THE BOARD

Nina D. Hoffman, SENIOR VICE PRESIDENT

PREPARED BY THE BOOK DIVISION

William R. Gray, VICE PRESIDENT AND DIRECTOR

Charles Kogod, ASSISTANT DIRECTOR

Barbara A. Payne, EDITORIAL DIRECTOR AND MANAGING EDITOR

David Griffin, DESIGN DIRECTOR

STAFF FOR THIS BOOK

Kevin Mulroy, EDITOR

Charles Kogod, ILLUSTRATIONS EDITOR

Michael J. Walsh, ART DIRECTOR

Anne E. Withers, RESEARCHER

Martha C. Christian, CONSULTING EDITOR

Carl Mehler, DIRECTOR OF MAPS

Kevin G. Craig, ASSISTANT EDITOR

Jehan Aziz, MAP PRODUCTION

Richard S. Wain, PRODUCTION PROJECT MANAGER

Meredith C. Wilcox, ILLUSTRATIONS ASSISTANT

Peggy J. Candore, STAFF ASSISTANT

MANUFACTURING AND QUALITY CONTROL

George V. White, DIRECTOR

John T. Dunn, ASSOCIATE DIRECTOR

Polly P. Tompkins, EXECUTIVE ASSISTANT

Mark A. Wentling, INDEXER